HYDROLOGICAL APPLICATIONS OF GIS

ADVANCES IN HYDROLOGICAL PROCESSES

Advances in Hydrological Processes is a series of books devoted to the publication of selected, original research and peer reviewed papers previously published in the Journal *Hydrological Processes.*

Series Editors   Malcolm G. Anderson
Norman E. Peters
Des Walling

Terrain Analysis and Distributed Modelling in Hydrology
Edited by K. J. Beven and I. D. Moore

Scale Issues in Hydrological Modelling
Edited by J. D. Kalma and M. Sivapalan

**Freshwater Ecosystems and Climate Change in North America
A Regional Assessment**
Edited by C. E. Cushing

Water Quality Trends and Geochemical Mass Balance
Edited by N. E. Peters, O. P. Bricker and M. M. Kennedy

**Distributed Hydrological Modelling
Applications of the TOPMODEL Concept**
Edited by K. J. Beven

Glacier Hydrology and Hydrochemistry
Edited by M. Sharp, K. S. Richards and M. Tranter

High Resolution Flow Modelling in Hydrology and Geomorphology
Edited by P. D. Bates and S. N. Lane

Hydrological Applications of GIS
Edited by A. M. Gurnell and D. R. Montgomery

HYDROLOGICAL APPLICATIONS OF GIS

Edited by

A. M. GURNELL

School of Geography and Environmental Sciences, University of Birmingham, UK

and

D. R. MONTGOMERY

Department of Geological Sciences, University of Washington, Seattle, USA

JOHN WILEY AND SONS

Chichester · New York · Weinheim · Brisbane · Singapore · Toronto

The papers in this volume were originally published in
Hydrological Processes – An International Journal, part 12(6): 821–994 (1998).

Reprinted December 2000

Other Wiley Editorial Offices

John Wiley & Sons, Inc., 605 Third Avenue,
New York, NY 10158-0012, USA

WILEY–VCH Verlag GmbH, Pappelallee 3,
D-69469 Weinheim, Germany

Jacaranda Wiley Ltd, 33 Park Road, Milton,
Queensland 4064, Australia

John Wiley & Sons (Asia) Pte Ltd, 2 Clementi Loop #02–01,
Jin Xing Distripark, Singapore 129809

John Wiley & Sons (Canada) Ltd, 22 Worcester Road,
Rexdale, Ontario M9W 1L1, Canada

Library of Congress Cataloging in Publication Data

Hydrological applications of GIS / edited by A.M. Gurnell and D.R. Montgomery.
    p. cm. – (Advances in hydrological processes)
    Includes bibliographical references.
    ISBN 0 471 89876 7 (alk. paper)
    1. Geographic information systems.   2. Hydrology – Data processing.
    I. Gurnell, A. M. (Angela M.) II. Montgomery, David R. III Series.

G70.212.H94   1999
551.48–d21
                                                    99–048239

British Library Cataloguing in Publication Data

A catalogue record for this book is available from the British
Library

ISBN 0-471-898767

Typeset in 10/12 Times
Printed and bound in Great Britain by Antony Rowe Ltd.

CONTENTS

Introduction: Hydrological Application of GIS; HYP, Vol. 12 (6); p. 821–824   1
A. Gurnell and D. Montgomery

1   Putting water in its place: a perspective on GIS in hydrology and water management;
HYP, Vol. 12 (6); p. 823–834   3
M. J. Clark

2   Data and databases for decision support; HYP, Vol. 12 (6); p. 835–842   15
A. M. Roberts and R. V. Moore

3   The treatment of flat areas and depressions in automated drainage analysis of raster
digital elevation models; HYP, Vol. 12 (6); p. 843–856   23
L. W. Martz and J. Garbrecht

4   A phenomenon-based approach to upslope contributing area and depressions in DEMs;
HYP, Vol. 12 (6), p. 857–872   37
W. Rieger

5   Large scale distributed modelling and the utility of detailed ground data
HYP. Vol. 12 (6): p. 873–888   53
F. G. R. Watson, R. B. Grayson, R. A. Vertessy and T. A. McMahon

6   Application of a GIS-based distributed hydrology model for prediction of forest harvest
effects on peak streamflow in the Pacific Northwest; HYP, Vol. 12 (6); p. 889–904   69
P. Storck, L. Bowling, P. Wetherbee and D. Lettenmaier

7   Modelling runoff and sediment transport in catchments using GIS; HYP,
Vol. 12 (6); p. 905–922   85
A. P. J. de Roo

8   Deciphering large landslides: linking hydrologic, groundwater, and slope-stability
models through GIS; HYP, Vol. 12 (6); p. 924–942   103
D. J. Miller and J. Sias

9   Regional test of a model for shallow landsliding; HYP, Vol. 12 (6); 943–956   123
D. R. Montgomery, K. Sullivan and H. Greenberg

10 Regional-scale assessment of non-point source groundwater contamination;
    HYP, Vol. 12 (6), p. 957–966   137
    K. Loague and D. L. Corwin

11 Synoptic views of sediment plumes and coastal geography of the Santa Barbara Channel,
    California; HYP, Vol. 12 (6); p. 967–980   147
    *L. A. K. Mertes, M. Hickman, B. Waltenberger, A. L. Bortman, E. Inlander, C. McKenzie
    and J. Dvorsky*

12 Morphological and ecological change on a meander bend: the role of hydrological
    processes and the application of GIS; HYP, Vol. 12 (6); p. 981–993   161
    A. M. Gurnell, M. Bickerton, P. G. Angold, D. Bell, I. Morrissey, G. E. Petts and J. Sadler

    Index   175

INTRODUCTION

Over the last two decades, the dramatic increase in the computer power available to hydrologists has led to significant developments in the way that hydrological research and operations are conducted. This volume, which is based on papers from a special issue of the journal *Hydrological Processes*, focuses on one area of such developments, the applications of GIS (geographical information systems) to the solution of hydrological problems. Over this period, GIS applications in environmental modelling have proliferated to take advantage of the spatial data representation capabilities of linking GIS systems to process-based models, and the examples presented in this volume represent many different levels of coupling between the GIS and the hydrological models employed to solve particular problems.

Hydrological applications of GIS are extremely varied. Whilst hydrological scientists have progressed in their representations of hydrological processes from lumped through semi-distributed to distributed hydrological models, water resource managers have followed a parallel route in the increasing spatial resolution with which assets, particularly infrastructure, have been represented, interrelated and managed. Common to both the research and management arenas is that the desire for increasing spatial resolution makes it attractive to work within a GIS framework. These developments are considered by Clark who highlights some of the technical and ethical ramifications of data quality and, in particular, of increasing spatial resolution. Furthermore, varied hydrological applications can be driven by different users accessing the same pool of information. As a result, the structure of the database that supports the GIS, the quality of the data and the way in which the database is managed lie at the heart of the development of many GIS applications. Roberts and Moore elaborate on some of these issues in the context of the development of a multidisciplinary database to support the United Kingdom's Natural Environment Research Council (NERC) Land Ocean Interaction Study (LOIS).

With the increasing availability of high-resolution digital elevation models (DEMs), the most widespread application of GIS in hydrology is the identification of drainage pathways and runoff contributing areas based on topographic form, and their coupling with hydrological models. Such applications frequently face problems concerning the accurate description of terrain. A fundamental issue is how to deal with topographic depressions and flat areas which may or may not be artifacts of the DEM. Chapters by Martz and Garbrecht and by Rieger review causes of such problems and propose some solutions. Chapters by Watson *et al.*, Storck *et al.* and de Roo examine the use of distributed hydrological models within GIS applications that address, respectively, the impact of precipitation patterns and the representation of vegetation and topography on estimated water yield; the impact of forest harvest on peak stream flow; and the estimation of catchment-scale soil erosion and sediment yield. The chapter by de Roo further develops the theme of coupling GIS to process-based models by illustrating the way in which runoff and soil erosion models can be both loosely coupled and embedded within a GIS.

Although catchment-scale hydrological modelling represents an important GIS application within hydrology, GIS has relevance to the solution of many other hydrological problems at local, catchment and regional scales. Other hydrological applications of GIS presented in this volume include the modelling of slope stability and landslide activity at both the site (Miller and Sias) and regional (Montgomery *et al.*) scales; the regional-scale assessment of non-point source groundwater contamination (Loague and Corwin); and an appraisal of the factors controlling coastal sediment plumes (Mertes *et al.*).

One of the great attractions of GIS is that it provides a framework for integrating data from disparate sources. The chapters in this volume illustrate integration of raster data from remotely sensed sources and fine DEM grids; vector data derived from contemporary and historical map and air photograph sources; and point data from hydrological and ecological monitoring networks. GIS permits the complex representation of these data and their modelling in an intuitive form that is capable of highlighting properties of the data that might otherwise remain unidentified. However, the power of GIS as a data integrating, manipulating and visualizing tool should not be allowed to disguise the dependence of the output on the accuracy of the input data and the potential of data manipulation to propagate and magnify errors. This is highlighted by Gurnell *et al.*'s focus upon the inherent errors associated with historical map and air photograph sources, and errors associated with their transcription through digitizing, to identify error limits that discriminate between genuine river channel change and data error. Watson *et al.* also focus on the manipulation of data and illustrate the sensitivity of estimated water yields to the spatial parameterization of precipitation, vegetation and topography. They show that an increase in the complexity of spatial parameterization does not necessarily result in better predictions of hydrological response. While GIS provides a rapidly expanding tool-kit that will allow hydrological scientists to address new questions, the development of such applications will inevitably raise a variety of interesting new problems.

This volume does not aim to be comprehensive, but the included chapters consider or illustrate some of the key issues that are currently relevant to hydrological applications of GIS.

Angela Gurnell and David Montgomery
Editors

1

PUTTING WATER IN ITS PLACE: A PERSPECTIVE ON GIS IN HYDROLOGY AND WATER MANAGEMENT

MICHAEL J. CLARK*

Geography Department and GeoData Institute, University of Southampton, Southampton, UK

ABSTRACT

The use of GIS (geographical information systems) in hydrology and water management has its roots in ideas about the relationship between climate, catchment, channel and society that emerged more than a hundred years ago. From these beginnings, hydrological GIS has come to be defined primarily by modelling in the science domain and by asset (notably infrastructure) management in the water and river management domain. In both contexts it can be demonstrated that data quality represents the ultimate constraint, but that the quest for higher resolution may carry with it some significant problems. These constraints are developed through an examination of the potential use of high resolution spatial data in flood insurance applications of GIS. While the issues raised have clear technical implications, they also have important professional and ethical ramifications which are worthy of consideration as a backdrop to the current and future status of GIS in hydrology and water management.

KEY WORDS   GIS; hydrological modelling; water management; information systems; flood insurance; ethics

THE ROOTS OF HYDROLOGICAL GIS

... The principal valleys in almost every great hydrographical basin in the world are of a shape and magnitude which imply that they have been due to other causes besides the mere excavating power of rivers.

A Manual of Elementary Geology, Charles Lyell (1852)

If it be admitted that the little stream has worn out the gutter in which it runs, it is hard to deny that the larger stream has not done similar work on a larger scale. The whole affair is indeed a mere question of time.

Physiography, T. H. Huxley (1880)

In articulating their contrasted views of the fluvial system, Lyell (1852) and Huxley (1880) showed fundamentally different perspectives on the role of the river. Placed together, their two views indicate that at some point soon after the middle of the nineteenth century the weight of informed opinion began to turn towards regarding rivers as creating, not just occupying, the valleys within which they flowed. Henceforth, rivers were deemed to have a place — their own place, of their own making — and the essential seed was sown for the slow and irregular development of the modes of thinking that now underpin GIS (geographical information systems) in hydrology. It would, perhaps, be premature to suggest that through this evolution the geographical approach to hydrology and rivers had been born, but it had certainly been conceived, and it is

---

* Correspondence to: Michael J. Clark, Geography Department and GeoData Institute, University of Southampton, Southampton SO17 1BJ, UK.

in this gestation that we must seek the roots of GIS in hydrology and water management. By considering science and management as distinct strands, it is possible to explore both the range of GIS applications and some of the shared challenges that the technology presents to academic and other professional users.

Against a background of such historical notions, Petts (1995) attempts to identify the essence of the 'modern' geographical approach to channel hydrology, and focuses on four conceptual building blocks. He suggests that the first foundation was laid by Strahler (1952), who asserted the reliance of fluvial geomorphology on the basic principles of mechanics and fluid dynamics, both involving spatially distributed parameters. The long-term response of channel systems to change in hydrological controls was introduced as a second focus through studies such as those of Leopold and Miller (1954) and Schumm (1968). The latter also introduced the notion of thresholds for channel change — an aspect which relates closely to spatial patterns. Thirdly, Schumm and Lichty (1965) carried these arguments forward with the recognition that cause–effect relationships could be modified or even reversed with variation in spatial and temporal scale, echoing Wolman and Miller's (1960) emphasis on the frequency and magnitude of geomorphological processes. Finally, and not surprisingly, these arguments were shadowed by more general debates on the emerging possibility of applying general systems theory within geomorphology (Chorley, 1962: Chorley and Kennedy, 1971).

Although more recent work on non-linear systems, scale dependency and spatial/temporal lag effects all serve to complicate relationships, it is clear that these four components identified by Petts (1995) create a basis for proposing that hydrology, catchment and fluvial systems interact in ways that might be addressed through GIS. Most obviously, and to many people most attractively, the relationship between hydrology, catchment and river output underpins the development of the distributed, physically based models that Brown (1995) sees as the core focus of GIS in hydrology, quoting Abrahart et al.'s (1994) defining assertion that hydrological GIS represents a modelling opportunity. While recognizing the widespread application of GIS within the corporate environment, largely for asset management purposes, Brown (1995) unambiguously views the hydrological focus as lying within the modelling function, with sampling, measurement, scale and accuracy representing key issues to be addressed. This stance attracts widespread consensus, and inevitably represents the starting point for any evaluation of the status of GIS in hydrology. Nevertheless, other viewpoints have been proposed — amongst them, Clark's (1995) proposal that

> ... the strategy that is being suggested lies in a progressive migration of GIS from an input role through one of support to full management functions. The strategy applies equally well in a decision context ... and in an operational context ... — both of which are core components in river management agencies.

These different perspectives are not in contention. Rather, it appears that within the broad domain of hydrology and related sciences, the priorities of GIS application will tend to depend upon context. Science has one set of targets, within which modelling occupies a key position. Management has a different priority, and draws upon GIS to underpin its decisions and operations. The distinction is one of balance, not exclusivity, and a diligent search reveals a wide spectrum of GIS use on both sides of the academic/professional watershed — with process-intensive (modelling) and data-intensive (archival; management)

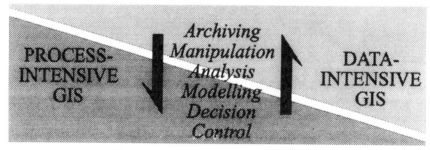

Figure 1. A concept of polarized GIS domains

modes (Figure 1) occupying overlapping positions in a seamless spectrum (Clark and Gurnell, 1991). Indeed, so important has this common ground between the apparently distinct scientific and management components of GIS become, that the term hydroGIS has been introduced to provide a shared focus for the hybrid technology and its users (Kovar and Nachtnebel, 1993).

The challenge faced by this paper is to identify the trends and issues that colour these different strands of GIS in hydrology so as to broaden the scope of this review beyond the conventional core. Nevertheless, these core topics provide a necessary and illuminating starting point, defined first by modelling in the hydrological science domain and then by asset management in the water and river management domain. In both contexts it will be suggested that data quality represents the ultimate constraint, but that the seemingly endless quest for higher resolution may carry with it some significant problems. These constraints are developed through an examination of the potential use of high resolution spatial data in flood insurance applications. While the issues raised have clear technical implications, they also have important professional/ethical ramifications which are worthy of consideration as a backdrop to the current and future status of GIS in hydrology and water management.

A FOCUS ON GIS MODELLING

Modelling draws to greater or lesser degree upon the geographic tradition (already discussed) that hydrology, catchment and fluvial systems interact closely and causally in time and space. Although the elements of hydrological modelling predate GIS by more than a century (Maidment, 1993), the two have converged strongly over the last 20 years. The current intensity of interest in the use of GIS for modelling is easily demonstrated. Almost 60% of the 152 papers in the 1993 and 1996 HydroGIS conference proceedings (Kovar and Nachtnebel, 1993, 1996) include the terms modelling, simulation or forecasting in the title and many others were clearly addressing components of these topics. No other aspect of GIS attracts such attention. Chow *et al.* (1988) offer a taxonomy of hydrological models based on randomness (deterministic/stochastic), spatial variation (lumped/distributed; space independent/space dependent) and time variation (steady flow/unsteady flow; time independent/time correlated) — thereby drawing attention to the pivotal position of the spatial dimension. The potential role of GIS to handle this dimension in environmental modelling has been cogently summarized by Goodchild (1993) as including

(i) Preprocessing data into a form suitable for analysis (scale, coordinate system, data structure, data model, etc.).

(ii) Direct support for modelling, so that tasks such as analysis, calibration and prediction are carried out by the GIS itself.

(iii) Post-processing data through reformatting, tabulation, mapping and report generation.

These three stages are not limiting, but they are representative of an overall range that has been widely recognized. They also provide a convenient structure within which it is readily apparent that the central role of the GIS in the actual process of modelling may vary from total to negligible. In the latter case, the simulation or prediction will be handled by a non-GIS hydrological model that is coupled to the GIS data input and output either 'loosely' (through the implementation of conversion protocols for data transfer) or 'closely' (where the GIS and model share a common data structure and both interact with the same database). This set of possible relationships, which has direct counterparts in most other sectors of the GIS domain, including asset management, is explored by Maidment (1993) to suggest several possible applications of the linkage.

(i) Hydrological assessment to represent hazard or vulnerability (through weighted and summed influences of significant factors rather than through physical laws).

(ii) Hydrological parameter determination, whereby the GIS provides inputs to the model in terms of parameters such as surface slope, channel length, land use and soil characteristics.

(iii) Hydrological modelling within the GIS, which Maidment sees as feasible provided that time snapshots or temporal averages are involved, not time-series.

(iv) Linking the GIS and hydrological models to utilise the GIS as an input and display device, including real-time process monitoring if the necessary (remotely sensed?) observations are available. The primary challenge remains the disparity between GIS and hydrological data models.

It is tempting to regard such categories as representing a spectrum of increasing significance of the role of GIS, with full linkage to modelling being seen as the ultimate expression of GIS contribution. However, such a set of evaluation criteria would almost certainly be misleading, in that the value of a GIS is not necessarily determined by the extent to which it is 'in control' of the processes concerned. Rigid views concerning the role of GIS can also be detrimental, in that overplaying its role can lead to failed expectations and thus a disillusioned backlash by disappointed users. Indeed, it is not even possible to view GIS as having a monopoly on spatial modelling, as is demonstrated by Howard's (1996) simulation of floodplain channel patterns through non-GIS techniques. This is just one of many ways in which hydrologists are playing out significant debates that rage with equal intensity in other branches of GIS.

GIS APPLICATIONS IN THE WATER INDUSTRY

While scientists were developing GIS in an essentially modelling dominated context, their counterparts in the water industry were facing related concerns about the role of GIS. In this more commercial domain of the water utilities, there developed, through the 1980s, a strong campaign (by power users as well as vendors) to position GIS as the core integrating technology — central to the information handling process and thus central to corporate success. Other applications were seen as satellites to the core GIS, and the promotion of GIS thus took on an almost Messianic quality (Figure 2). The early 1990s were also characterized by a GIS-centric approach which placed GIS as the key facilitating technology for corporate re-engineering.

Maffini (1997) argues conversely that the database has already replaced the GIS as the integrating core of corporate information handling, and suggests that this position will itself shortly be usurped by network computing capability (see Figure 2). It seems that the perceived architecture of information processing (and the place of GIS within it) is highly dependent upon technological drift. While it can certainly be argued that GIS integrates elements as varied as automated mapping (AM), facilities management (FM), remote sensing, land information systems (LIS) and spatial statistics, the notion of its being the 'central' technology is somewhat misleading. Increasingly, in practice, GIS serves primarily as an input to management information systems (MIS) in the corporate domain and modelling/control systems in the research domain (Figure 3). For example, at North West Water, Britain's second largest water utility, the GIS user community covers more than 350 engineers, schedulers, planners, project managers and customer service personnel (Fitzgibbon, 1996). GIS manages information on the physical infrastructure (for water supply and sewage) and on its operations, including development and maintenance. When it is realized that a GIS application at full enterprise scale could have thousands of users, hundreds of concurrent users, tens of millions of records and hundreds of thousands of daily transactions, it becomes clear why database operations are the key to successful implementation. In water utilities, about 70–80% of all corporate information is tied to geography: customers have a location as do business transactions, orders, billings, operations and infrastructure management. All the foundations of business performance have location, and thus benefit when location is built into planning, design, operation and maintenance through the use of GIS (Clark and Maffini, 1997).

While there is some logic in such a focus on GIS *per se*, as both an information system gateway and an integrator, the resulting overstatement of GIS capability led to a history of project underperformance in the 1980s. More recently, there has been a welcome shift towards regarding GIS as one of several application domains that encircle the core technology, which some would now identify as being the database. At the present time, it is probably more than coincidental that corporate commentators pay so much attention to the database as the key to the management information system (MIS) in the same way that Maidment (1993)

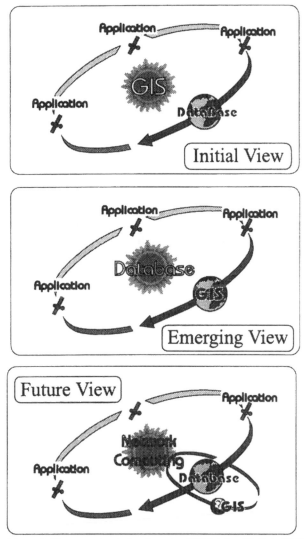

Figure 2. Three constructs for the role of GIS (after Maffini, 1997)

focuses on the data model as the key to functionally efficient GIS modelling. Effective data acquisition and modelling may be seen to lie close to the heart of hydrological model enhancement, as Maidment (1993) concludes

> It is probably true that the factor most limiting hydrologic modelling is not the ability to characterize hydrologic processes mathematically, or to solve the resulting equations, but rather the ability to specify values of the model parameters representing the flow environment accurately. GIS will help overcome that limitation.

Such a perspective maintains modelling as a priority focus for GIS in hydrology, but this brief review of its role has indicated that hydrological GIS and generic GIS share more characteristics and professional traits than might have been expected. The input–model–output structure of conventional GIS is paralleled by the Input-Transaction-Output structure of a water utility asset management GIS (Figure 4), and the handling of transaction processing can be seen as the same order of technical challenge as the handling of a modelling

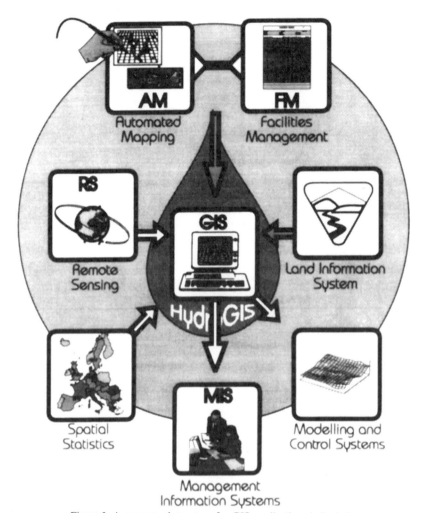

Figure 3. A conceptual structure for GIS applications in hydrology

algorithm. It is therefore appropriate to broaden the argument in order to seek this more general perspective by reviewing the implications of GIS data-handling in the scientific and management contexts.

DATA POTENTIAL AND CONSTRAINT IN HYDROLOGICAL GIS

Data quality is a prime constraint on the power and reliability of the hydrological application of GIS in both the scientific and commercial contexts, and relates in large measure to precision, accuracy and error. At the most fundamental level, precision and accuracy are often unavoidable prerequisites to building a successful model of a hydrological system or its outcome in the fluvial system. For example, Gurnell *et al.* (1994) and Downward (1995) explore the roots of error in a GIS-based approach to deriving and analysing planform changes in river channels from historical mapping. Gurnell (1997) extends related arguments to channel attribute derivation from air photography. These studies highlight the distinction between inherent errors in the specification of geographical space and operational errors in the acquisition of data. Significant precision is clearly achievable, but many studies support the well-known conundrum that in the presence of systematic errors it is entirely possible to achieve measurements that are precise but not accurate, and *vice versa* (Thapa and Bossla, 1992).

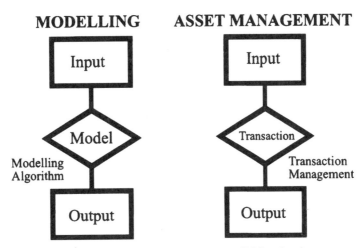

Figure 4. Modelling and asset management as parallel functional structures

Bates *et al.* (1996) pursue the implications of resolution into hydraulic modelling *per se*. In a test on an 11-km reach of the River Culm, they compare models based on finite element meshes with 2040 and 9600 linear triangular meshes respectively, both meshes being interpolated from elevation data from 1:25 000 scale maps supplemented by field-derived cross-sections. A subsequent test of a 12-km reach of the River Stour employed 4396 and 11 265 computational elements, respectively, this time derived from a 10×10 m DTM supplied by the UK Ordnance Survey, again supplemented by field-surveyed cross-sections. The comparisons suggested that the higher resolutions provided improved topographic parameterization which was able to depict a greater degree of spatial variability, and on this basis the authors conclude that high space–time resolution modelling is becoming data limited. They draw attention to the progressive loss of information in the passage from real-world topography through DTM representation to model discretization, but nevertheless debate the possibility that the optimum relationship between process scale and mesh resolution for GIS-based hydraulic modelling has yet to be determined.

However, the constant quest to minimize error by increasing resolution is not without its detrimental implications, and high resolution has long been seen as a holy grail with a hidden backlash. At least four significant challenges accompany high resolution and render it problematic.

1. Both spatial and temporal resolution carry an overhead of increasing data volume, with concomitant increases in storage and processing speed. However, in principle, and to a growing extent in practice, technological advances tend to annul this difficulty.

2. The remote sensing community working with a pixellated raster data model of the Earth's surface has long recognized that higher resolution increases the scene noise whereby adjacent pixels carry different values in response to minor and local changes in ground surface properties, giving a speckled effect. Handling these data in applications such as models often requires that they be smoothed, which loses the advantage of the high resolution. The implications are explored by Woodcock and Strahler (1987).

3. Many of the features represented in the GIS model of the real world are by their nature fuzzy and time-bound transitions which are simplified to an increasingly unrealistic rigid and static spatial representation as resolution increases. Milton *et al.* (1995) quote the example of a mid-channel bar, which remains fuzzy whatever the resolution of depiction.

4. In a few high-profile applications — notably those associated with the insurance market — high resolution implies a specificity that decreases the unknowns or errors in estimating risk probabilities. This apparently beneficial attribute has important professional implications which may, ironically, render it a net detriment.

The debate concerning optimum scale and resolution shows no sign of exhaustion, and it may indeed be that there is no absolute optimum — merely a series of trade-offs between which the choice rests largely on the specific needs of an individual application. The drive towards higher resolution has been seen for some time to represent a conceptual as well as technical challenge (Burrough, 1989), but Hoover *et al.* (1991) identify the dangers of adopting an over-coarse resolution that is incapable of representing significant elements of the hydrological system. Clearly, there is a technical problem inherent in setting ideal resolution and precision, but it is also important to recognize the inherent managerial implications.

AN EXAMPLE OF APPLICATION: GIS AND FLOOD INSURANCE

The data resolution dilemma is nowhere more starkly faced than in the context of natural hazard (natural perils) insurance. The concept of insurance is based upon the classic notion that the burden of the few (those affected by the hazard event) shall fall lightly on the shoulders of the many (the client base paying premiums to the insurer). After several centuries of operation, this device continues to serve both individual and society extremely well. The individual is not only helped in remedying the immediate situation, but is also permitted to recover (physically and economically) much more rapidly than would be possible if unaided. Society is simultaneously advantaged by the restoration of economic and social fabric, and by the fact that the burden of recovery has been delegated. Not only is insurance effective, it is also presumptive — working increasingly to encourage action to reduce exposure to, and the impact of, natural hazards. Clearly, insurance is much more than just a commercial activity that can be left to prosper or decline according to market forces.

Flood insurance is typical of the natural hazards management sector in that both the market and the insurance cover provided have tended to develop spasmodically as a reaction to specific events (Arnell *et al.*, 1984). While the USA opted for a model involving state subsidy under the National Flood Insurance Act (1968), other flood insurance programmes remain essentially private commercial contracts between insured and insurer, and the premium structure has been related grossly to a perceived overall probability spectrum of risk based on limited past experience. The premium model is often extremely crude, so that, in principle, many people have *de facto* been overinsured or underinsured, but the model has operated with reasonable effectiveness in countries such as the UK.

In order to permit premium income to sustain claims payments, it is necessary *both* to employ an insurance premium rating model that properly reflects the probability of future flood claims, *and* to sell sufficient policies at the thus determined rates to fund the future profile of claims. Simplistically, the rating calculation will take the form either of a statistical model generated by claims history, or of a process model using understanding of the hazard/catastrophe concerned as a basis for prediction of location and temporal probability. The sale of policies will reflect customer awareness of the peril, coupled with a pricing policy that will be sensitive to adverse selection (an unwelcome situation whereby only those at greatest risk actually purchase insurance). The problem of creating a rating model is exacerbated when the natural process is cumulative in effect over a long time period, when the process is subject to long-term intrinsic change (which is believed to apply to most climatically induced perils) and when human intervention in the natural system or its effects begins or changes. In all such cases, past claims history will be a misleading indicator of future performance, and process modelling should, in principle, be preferred, with significant implications for GIS.

Herein lies a fundamental dilemma. Accurate rating demands accurate data, whether on claims experience or causative process. High resolution techniques and data sets exist, or could be developed, that would greatly improve the ability to estimate the flood risk actually associated with a particular person, property or event. Incorporated into a GIS that includes client addresses and three-dimensional property location (already in wide use amongst insurers and re-insurers), this information should allow a high precision calculation of appropriate premium, thereby, conceivably, deeming some people uninsurable, and rendering others uninsurable at a price acceptable to them. The greater the resolution of the data, the greater the tendency for the insurance underwriter to remove high risk individuals from the insurance market. By the same token, low risk individuals will increasingly either withdraw from cover voluntarily, or will become actively opposed to a premium structure that fails fully to reflect their relatively low risk. Ultimately, so few people would opt for cover at an economic rate that there is a real chance that the concept of the loss of the few falling upon the many would break down. It is, moreover, clear that the threat of client litigation on the basis that an underwriter has negligently misclassified a risk — with consequent detriment to property value or usage — increases substantially with the specificity of the premium model.

There are signs that this position has already been reached with insurance against the effects (and clean-up costs) of land contamination. Retrospective and long-term gradual effects have already been removed from new policies, and even the continuing availability of cover for sudden events has an uncertain future. Exclusions are far-reaching, and the risk covered is often so tightly constrained that cover becomes unattractive to the purchaser and the total market is seen as very low. The underwriters' low understanding of the peril is leading to a high degree of caution which is suffocating both cover and market. Premiums are not closely linked to relative risk, so that risk management investment by the client is not reflected in premium level — a lost opportunity to provide an incentive for risk control. An immediate response will be for large buyers to assess risks so as to carry the high certainty perils in-house or with captive companies, leaving the insurance market with adversely selected, low certainty perils. Even more seriously, the environmental audit that is necessary to permit the landowner/operator to tackle the contamination peril will be seen as disadvantageous if the information revealed is used to downgrade the insurance status of the site. Paradoxically, insurance in this case is working against risk control and mitigation — the exact opposite of the normal and desired position — and with the advent of high resolution data the same could happen to the flood insurance market. On the basis of the foregoing argument, the environmental insurance dilemma appears to be essentially twofold.

1. High-resolution data on risk and peril are necessary in order to permit GIS applications to reduce the exposure of the underwriter and permit insurance cover to be offered economically; but, at the same time, precise information renders more individuals uninsurable, potentially increases adverse selection, encourages litigation and ultimately suffocates the market.

2. By using high resolution information on risk to downgrade the insurability of individual clients, the underwriter is providing a disincentive to remediation (which itself is dependent upon information generation), thus countering the mission of bodies such as the UK Loss Prevention Council.

The difficulty that is faced in creating viable GIS applications from routinely available commercial spatial data lies not so much in data error as in user misconception. For example, the UK Ordnance Survey (OS) is a world leader in the provision of accurate digital data, and is meticulous in documenting both resolution and error in its product catalogues so that there is no objective possibility of misunderstanding. Nevertheless, in 1997 media interest was suddenly focused on one OS data product called Land-Form PROFILE®, depicting it as new and tailor-made for the insurance industry

> A sophisticated 3-D map produced by Ordnance Survey shows which houses are likely to flood in a given street and is claimed to be highly accurate. (Insurance) price rises or even blacklisting are expected to follow for homeowners on the list.
>
> *Sunday Times, 1997*

Insurers using this data set are interested in flood height differences of tens of centimetres, but the height accuracy of the spatial data is quoted as 1·5 m root mean square error (Ordnance Survey, 1997). The OS documentation is careful to spell out in full the implications of rms error, making it entirely clear that 1% of points can be expected to lie more than 2·5 rms from the quoted height in the data set — an error in this case of $> \pm 3.75$ m! The mismatch between producer and user perceptions of accuracy is thought-provoking.

These problems would be potentially damaging even if the insurance industry's mission was exclusively one of business return on investment. They are doubly challenging given the fact that insurance has come to be a pivotal component of the national and international infrastructure through which advanced societies seek to manage risk and mitigate its detrimental effects. The predicament is both technical and philosophical, and the identification of a solution (albeit a partial or temporary solution) would be beneficial to the underwriter/re-insurer concerned and to society (individually and severally). It is clear that the challenges faced by GIS in hydrology extend well beyond the technical difficulties of close coupling of GIS and hydraulic models, despite the obvious value of this task.

PROFESSIONAL PROTOCOLS IN CONTEXT

Although flood insurance provides an effective indication of the way in which the implications of GIS in hydrology are not restricted to technical modelling issues, but rest equally in the application domain, the underlying issues are still more general. There has, in recent years, been a powerful backlash against the use of GIS, focusing on the fear that such information technologies are socially divisive and damaging (Pickles, 1995). There is a temptation amongst scientists to regard such attacks as intemperate, but it does not take much contemplation to confirm that GIS and related IT systems do indeed have enormous power, which is open to misuse and abuse. They transform understanding, decisions and operations. They revolutionize the way that employees work, and the way that businesses and professions relate to their customers. The example of flood insurance GIS is sufficient to demonstrate that GIS does make a huge and direct difference to everyone, and it *must* be handled with the greatest of professional respect. To this end, a professional code has been devised (Clark and Maffini, 1997) which could well serve to provide operating standards for GIS practitioners and users.

- Have you *actively* checked whether the data uses and processing that have been proposed meet corporate mission standards?

- Have you *actively* worked to ensure equitable and responsible access to information? There are legal safeguards, but you can usually go further and do better.

- Are your data documented in such a way that everyone using them is *bound* to be aware of their resolution and other limitations or constraints?

- Have you *ensured* that data processing techniques and protocols are so explicit that every user can assess their suitability and adequacy for the task in hand?

- Have you established an error-tracking audit to *ensure* that the system meets explicitly defined quality targets?

- Have you *personally* asked whether what you are doing is beneficial to the business, the customer and society? You cannot transfer this responsibility to someone else!

Data manipulation, analysis, modelling and decision support are all potential social interventions that relate to the use of GIS in hydrology and water management. We now begin to see something of their power, and it is timely to adopt approaches to ensure that this power is used beneficially (Clark, 1996). Grayson *et al.* (1993) note the ethical challenge of GIS visualizations of hydrological models concealing subtleties, and put the professional responsibility for alerting users to the capabilities, assumptions and constraints of the GIS directly on to the GIS practitioner. The code of practice presented above can go far towards establishing this professionalism.

A PERSPECTIVE ON GIS IN HYDROLOGY AND WATER MANAGEMENT

In this brief survey the emergence of GIS in hydrology has been traced from nineteenth century insights about the causal relationships between weather, hydrology and fluvial systems through to twentieth century technological devices for modelling such relationships. In professional terms, it has been demonstrated that water utilities adopt similar protocols to handle very different relationships, focusing on transaction processing of asset management interventions (in the broadest sense) in place of hydrological modelling.

Assessing the scope of the future of information technology is a high risk activity, but it does make possible a tentative evaluation of the priorities for GIS in hydrology and water management. In this respect, the future will be driven by a blend of technological, professional and social factors. Technologically, the effective acquisition and manipulation of high resolution spatial data will become commonplace, while the networked distribution of information and processing power to the individual will become the norm in the developed world. These trends will permit a higher degree of modelling and transaction processing (or simulation), such that both water science and water industry can expect significant increases in efficiency. Social benefits may accrue, both from this efficiency and from the competitiveness that it produces, but social dilemmas will remain in the GIS context. Limitless access to information may lead to real detriment as individuals find themselves targeted by seemingly omniscient but insensitive systems: sometimes it is better not to know every potential problem. Information élites will emerge to provide the twentyfirst century counterparts of twentieth century commercial and political élites. GIS will be strengthened, but progressively subsumed into modelling, management and operational control systems (much as spatial management is already subsumed into the flight control systems of modern aircraft), and codes of professional practice will become essential to arbitrate between the GIS practitioner and society.

The trends first glimpsed by Lyell and Huxley more than a hundred years ago have clarified and elaborated, but they have not simplified. The challenges that now emerge have social and professional ramifications that have hitherto been invisible to most hydrologists and GIS professionals, but they can no longer be ignored or trivialized. Only when both scientific rigour and robust professional ethics have been achieved can we regard GIS as being successful in its quest to put water in its place.

REFERENCES

Abrahart, R. J., Kirkby, M. J., and McMahon, M. L. 1994. 'MEDRUSH — a combined geographical information system and large scale distributed process model', in Fisher, P. (ed.), *Proceedings of GIS Research UK (GISRUK), Leicester, 1994*, pp. 67–75.

Arnell, N. W., Gurnell, A. M., and Clark, M. J. 1984. 'The role of crises in prompting institutional change in response to flood hazard: flood insurance in the UK', *Applied Geography*, **4**, 167–181.

Bates, P. D., Anderson, M. G., Price, D. A., Hardy, R. J., and Smith, C. N. 1996. 'Analysis and development of hydraulic models for floodplain flows', in Anderson. M. G., Walling, D. E. and Bates P. D. (eds), *Floodplain Processes*, John Wiley & Sons, Chichester. pp. 215–254.

Brown, T. J. 1995. 'The role of geographical information systems in hydrology', in Foster, I. Gurnell, A. M., and Petts, G. E. (eds), *Sediment and Water Quality in River Catchments*, John Wiley & Sons, Chichester, pp. 33–48.

Burrough, P. A. 1989. 'GIS — more than a spatial database', in *Managing Geographical Information Systems and Databases*. University of Lancaster. pp. 1–16.

Chorley, R. J. 1962. 'Geomorphology and general systems theory', *US Geological Survey Professional Paper, 500B*. USGS Reston, VA.

Chorley, R. J., and Kennedy, B. A. 1971. *Physical Geography: a Systems Approach*. Prentice Hall, London.

Chow, V. T., Maidment, D. R., and Mays, L. W. 1988. *Applied Hydrology*. McGraw-Hill, New York.

Clark, M. J. 1995. 'Information flow for channel management', in Gurnell A. M. and Petts G. E. (eds), *Changing River Channels*. John Wiley & Sons, Chichester. pp. 263–276.

Clark, M. J. 1996. 'Professional integrity and the social role of hydro-GIS', in Kovar, K. and Nachtnebel, H. P. (eds), *HydroGIS96: Application of Geographic Information Systems in Hydrology and Water Resources Management (Proceedings of the Vienna Conference, April 1996), IAHS Publ,*. **235**, 279–287.

Clark, M. J. and Gurnell, A. M. 1991. 'Integrating hydrological databases — GIS and other solutions', *Proceedings of Third National Hydrology Symposium, University of Southampton 16–18 September 1991, British Hydrological Society*. pp. 6·1–6·6.

Clark, M. J. and Maffini, G. 1997. *A Step-by-Step Guide to Enterprise GIS Decisions*. SHL VISION* Solutions, London. 28 pp.

Downward, S. 1995. 'Information from topographic survey', in Gurnell A. M. and Petts G. E. (eds), *Changing River Channels*. John Wiley & Sons, Chichester. pp. 303–323.

Fitzgibbon, P. 1996. 'Mapping matters: making information count', *Utility Week*, **20**, September 1996.

Goodchild, M. F. 1993. 'The state of GIS for environmental problem-solving', in Goodchild, M. F., Parks, B. O., and Steyaert, L. T. (eds), *Environmental Modeling with GIS*. Oxford University Press, Oxford. pp. 8–15.

Grayson, R. B., Blöschl, G., Barling, R. D., and Moore, I. D. 1993. 'Process, scale and constraints to hydrological modelling in GIS', in Kovar, K. and Nachtnebel, H. P. (eds), *HydroGIS93: Application of Geographic Information Systems in Hydrology and Water Resources Management (Proceedings of the Vienna Conference, April 1993), IAHS Publ.*, **211**, 83–92.

Gurnell, A. M. 1997. 'Channel change on the River Dee meanders, 1946–1992, from the analysis of air photographs', *Regul. Rivers, Res. Mgmt*, **13**, 13–26.

Gurnell, A. M., Downward, S. R., and Jones, R. 1994. 'Channel planform change on the River Dee meanders, 1876–1992', *Reg. Rivers, Res. Mgmt*, **9**, 187–204.

Hoover, K. A., Foley, M. G., and Heasler, P. G. 1991. 'Sub-grid-scale characterisation of channel lengths for use in catchment modeling', *Wat. Resourc. Res.*, **27**(11), 2865–2873.

Howard, A. D. 1996. 'Modelling channel evolution and floodplain morphology,' in Anderson. M. G., Walling, D. E., and Bates P. D. (eds), *Floodplain Processes*. John Wiley & Sons, Chichester. pp. 15–62.

Huxley, T. H. 1880. *Physiography: An Introduction to the Study of Nature*, 3rd edn. Macmillan and Co., London. 384 pp.

Kovar, K. and Nachtnebel, H. P. (eds) 1993. *HydroGIS93: Application of Geographic Information Systems in Hydrology and Water Resources Management (Proceedings of the Vienna Conference, April 1993)*, *IAHS Publ.* **211**, 693 pp.

Kovar, K. and Nachtnebel, H. P. (eds) 1996. *HydroGIS96: Application of Geographic Information Systems in Hydrology and Water Resources Management (Proceedings of the Vienna Conference, April 1996)*, *IAHS Publ.* **235**, 711 pp.

Leopold, L. B. and Miller, J. P. 1954. 'A postglacial chronology for some Alluvial valleys in Wyoming', *US Geological Survey Water Supply Paper*, **1261**, 1–90. USGS Reston, VA.

Lyell, C. 1852. *A Manual of Elementary Geology*, 4th edn, London. Quoted in Chorley, R. J., Dunn. A. J., and Beckinsale, R. P. 1964. *A History of the Study of Landforms*. Methuen & Co. and John Wiley & Sons, Chichester. pp. 1, 183.

Maffini, G. 1997. 'Using GIS technology to advance and optimise asset management', Presentation to *Strategic and Profitable Asset Management Conference*, London, 21 May 1997.

Maidment, D. R. 1993. 'GIS and hydrologic modeling', in Goodchild, M. F., Parks, B. O., and Steyaert, L. T. (eds), *Environmental Modeling with GIS*. Oxford University Press, Oxford. pp. 147–167.

Milton, E. J., Gilvear, D. J., and Hooper, I. D. 1995. 'Investigating change in fluvial systems using remotely sensed data', in Gurnell A. M. and Petts G. E. (eds), *Changing River Channels*. John Wiley & Sons, Chichester. pp. 277–302.

Ordnance Survey 1997. *Products and Services Catalogue 1997 for the Business and Professional User*. Ordnance Survey, Southampton. 53 pp.

Petts, G. E. 1995. 'Changing river channels: the geographical tradition', in Gurnell A. M. and Petts, G. E. (eds), *Changing River Channels*. John Wiley & Sons, Chichester. pp. 1–23.

Pickles, J. (ed.) 1995. *Ground Truth: the Social Implications of Geographic Information Systems*. The Guildford Press, London and New York. 248 pp.

Schumm, S. A. 1968. 'Speculations concerning palaeohydrologic controls of terrestrial sedimentation', *Bull. Geol. Soc. Am.*, **79**, 1573–1588.

Schumm, S. A. and Lichty, R. W. 1965. 'Time, space and causality in geomorphology', *Am. J. Sci.*, **263**, 110–119.

Strahler, A. N. 1952. 'Dynamic basis of geomorphology', *Bull. Geol. Soc. Am.*, **63**, 923–937.

Thapa, K. and Bossler, J. 1992. 'Accuracy of spatial data used in geographic information systems', *Photogramm. Engng Remote Sens.*, **58**, 835–841.

Woodcock, C. E. and Strahler, A. H. 1987. 'The factor of scale in remote sensing', *Remote Sens. Environ.*, **21**, 311–332.

Wolman, M. G. and Miller, J. P. 1960. 'Magnitude and frequency of forces in geomorphic processes', *J Geol.*, **68**, 54–74.

2

DATA AND DATABASES FOR DECISION SUPPORT

ANNE M. ROBERTS* AND ROGER V. MOORE

Institute of Hydrology, Crowmarsh Gifford, Wallingford, Oxfordshire, OX10 8BB, UK

ABSTRACT

Heightened awareness of the economic value of good quality multidisciplinary data sets, both to UK science and UK Ltd, has ensured that data management now has a high priority. As a result all new Natural Environment Research Council (NERC) Thematic Programmes must therefore have a data management strategy based on the NERC Data Policy. This paper describes the data management strategy for NERC's Land Ocean Interaction Study and the GIS tool used to help implement it in one of the programme's data centres. Many of the issues raised and addressed by this data management strategy are relevant to any major multidisciplinary project.

KEY WORDS   decision support systems; sustainable models; Land Ocean Interaction Study; data management; data model design; GIS tools

INTRODUCTION

The Land Ocean Interaction Study (LOIS) is the UK Natural Environment Research Council's (NERC) £30 million flagship thematic programme, which comes to its conclusion in April 1998. Its magnitude and the consequent volume and diversity of the data needed to support it have highlighted a number of major issues relating to the use and management of data. LOIS has attempted to address many of these, which are common to any multidisciplinary project. Of these issues, the key ones that this paper will discuss are:

- the volume and variety of the data sets;

- the harmonization of similar data across different disciplines of science or across organizations;

- the development of a data model that enables an holistic approach to the management of all environmental data;

- removing the need for specialist database systems and tools by incorporating their facilities within mainstream databases (Browne, 1995)

- the provision of access to data:

  (i) balancing the rights of the collector and the funding agency,

  (ii) observing copyrights and intellectual property rights (IPR),

  (iii) providing on-line meta-data catalogues,

  (iv) providing direct remote access and removing the need for staff to service,

  (v) requests from remote users,

  (vi) improving the user interface;

---

* Correspondence to: Anne M. Roberts, Institute of Hydrology, Crowmarsh Gifford, Wallingford, Oxfordshire, OX10 8BB.

- the need for quality assurance (QA) and an audit trail of change designed for users as opposed to database administrators; and

- the need for effective quality control (QC) of data.

Returning to the LOIS programme as a whole, its objectives are to

 (i) understand the flux movements between the land, the sea and the atmosphere, and

(ii) develop models of coastal zone processes.

Thus, LOIS should leave the policy makers and planners in a position to develop operational decision support systems to help them achieve the sustainable management practices that were one of the ideals of the Rio Earth Summit (HMSO, 1994).

LOIS consists of five integrated subcomponent studies, which are: the River, Atmosphere and Coastal Study (RACS); the Shelf Edge Study (SES); the North Sea Modelling Study (NORMS); the Land Ocean Evolution Perspective Study (LOEPS); and the DATA component. Consortia of university and NERC institute research scientists have been brought together to work on problems mainly within the traditional environmental disciplines of rivers, atmosphere, coast, geology and the deep sea. However, the coastal zone operates as an integrated whole. What sets LOIS apart and ensures its relevance to the policy makers and planners, is that a significant part of the programme is directed towards understanding processes that cross the disciplinary divides.

Cross-disciplinary research is necessary because the decision making processes that LOIS seeks to support are no longer considering single issues in isolation set against a steady state background. Today's planners must attempt to consider all the ramifications of a given decision against a whole range of possible scenarios. The reasons for this change are twofold, the first being rather neatly summarized by Jamieson and Fedra (1996):

> While resource utilisation remains small, interactions between these (water supply, land drainage, hydropower generation, effluent disposal, recreation and amenity) different interests are largely absorbed by the natural buffering within the physical system. However, as demands increase, it soon becomes necessary to co-ordinate activities. Eventually, there comes a time when, to realise the full potential, the only sensible way to proceed is to consider the basin as one complex integrated system.

The second is that we are now almost certainly in an era of climatic change. The assumptions of climatic stability were probably never valid, but in the past neither the data nor the computing power existed to enable any other assumption.

The implication of this situation is that, the decision support systems for use at the planning level will need to model, not just a single process in isolation, but many processes and their interactions. This in turn will require major advances in the design of supporting database systems, if the whole system is not to fail owing to the variety of data that must be gathered from a host of disparate databases spread across many locations.

Most of LOIS has been devoted to the development of chemical, hydrological, biological and marine process models. These process models will enable the investigation of possible decisions set against different background scenarios. In the research phase, the models have drawn their data from a variety of databases and 'ad hoc' file structures. However, this will not be a viable option once they are included in operational systems. Therefore, either a common interface to these different databases, or a database system capable of holding all environmental data is required. In this context, a common interface means a query language (or equivalent mechanism), together with a logical data model in which all data may be perceived to be held. An obvious example is the query language SQL and the two-dimensional tables used in relational databases.

Unfortunately, at the time of writing, SQL does not readily allow the formulation of questions that span both space and time. LOIS has therefore expended considerable effort and resources in developing and implementing a generic database design for storing coastal zone information. In parallel, significant strides have been made in developing the organization necessary to assemble and disseminate data for major thematic programmes such as LOIS. This paper will explain the lessons learnt, the tools used and will identify where future work is required.

DATA MANAGEMENT IN LOIS

Past NERC thematic programmes have left scientists to acquire and manage their own data sets. Declining government funding and the need to find alternative sources of revenue have alerted NERC to the value of data and the need to exploit them properly. Lowry *et al.*, in their papers, examine the problems of data management and quality control to support thematic programmes in the marine sciences (Lowry and Cramer, 1995; Lowry and Loch, 1995). They found that the scale and cost of the science involved has raised awareness that the data sets produced are a valuable long-term resource, and that by leaving data distributed throughout the organization an expensive and valuable asset was rendered unexploitable. The consequence of this has been the development of a NERC Data Policy (NERC, 1996) and the provision of NERC Designated Data Centres to hold important data sets.

In relation to LOIS, the key elements of the policy are that:

- data collected with LOIS funding are available to all working in the LOIS Programme;
- data must be submitted to a LOIS Data Centre;
- users of data must inform and acknowledge originators;
- copyright must be observed;
- all LOIS data are to be made available to bona fide researchers at cost; and
- data may be sold for commercial applications at market prices.

As a result of this policy, there are considerable challenges for the data centres in the form of:

- the diversity of data;
- the volume of data;
- the lack of standards;
- the number of suppliers;
- ensuring IPR/copyright are honoured;
- maintaining QA and audit trails;
- maintaining security and confidentiality;
- finite resources; and
- the demand for data by a geographically dispersed user base with widely varying computing skills.

These challenges have been met by developing organizational structures and new systems.

ORGANIZATION

Like most scientific programmes, LOIS is managed by a steering committee, which consists of a mixture of scientists from NERC institutes, and UK universities and includes a representative from each of the LOIS subcomponents. The steering committee confines itself to directing the science policy and major decisions on the allocation of funds. It is the task of the subcomponent chairmen of LOIS to translate policy into action.

In respect of the DATA component, this has been achieved through the establishment of five LOIS data centres:

- the LOEPS Data Centre at the British Geological Survey, Nottingham;
- the RACS River Data Centre at the Institute of Hydrology (IH), Wallingford;
- the RACS Coastal Data Centre at the Institute of Terrestrial Ecology, Huntingdon;
- the RACS Atmosphere Data Centre at University of East Anglia, Norwich; and
- the SES and NORMS Data Centre at the British Oceanography Data Centre, Bidston.

Their common objectives are to:

- identify user needs;
- acquire major data sets from various sources;
- provide data management;
- disseminate data to users;
- set format standards;
- ensure long-term security; and
- manage a budget for the acquisition/dissemination of identified data sets.

The activities of the data centres are coordinated by the LOIS DATA Committee. Its role is to set standards, purchase cross-component data sets and manage cross-data centre tasks. Examples of these tasks include, the production of a CD-ROM of LOIS data and coordinating requests to and from external agencies. This particular problem has been overcome by nominating a data centre as the contact point for each major supplier and, by obtaining data in bulk transfers, thus eliminating large numbers of requests for small quantities of data by individuals.

A further objective of the LOIS DATA Committee is to develop techniques for the integrated management and distribution of environmental data (NERC, 1992). In the case of the Rivers Data Centre at IH, the main task has been to assemble hydrological, water quality, water quantity and land use data together with many other supporting data sets. These data vary in space and time, come from many suppliers and have not been collected to any common standards. However, the scientists need a single harmonized data set covering the area of the east coast of Scotland and England, from the Tweed to Lowestoft. To this end, the Rivers Data Centre has used for its database activities the water information system (WIS), which provides a single storage system within which all these disparate data types can be held and visualized. The remainder of this paper will describe the basic concepts within the WIS design that have been developed in order to pursue the holistic approach to data management that LOIS requires.

THE WIS LOGICAL DATA MODEL — THE WIS CUBE

At the user level, the world as perceived by WIS is composed of features. These are any objects whose description in space and time the user wishes to record. It is the user who decides and defines the types of feature for which data are to be held. The descriptions of features and the events observed at them are recorded in terms of attributes. A wide range of spatial and non-spatial data types are supported, which allows WIS to record most types of attribute information likely to be of interest to LOIS scientists.

Traditional database design, which is ably described by Collet *et al.* (1996), tends to lead to highly specific solutions as shown in the simplified extract of the design of the GESREAU conceptual data model. This specific solution is quite acceptable for a stable situation where new data types are not being regularly introduced. In the LOIS Programme, and in many other situations where GIS solutions are needed, this is not the case. Specific solutions have a further disadvantage in that they make it very much harder to design

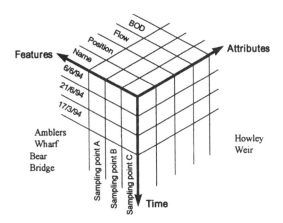

Figure 1. The WIS cube

supporting generic query systems. The WIS system has therefore taken the design process a step further to produce a genuine generic conceptual model for the input, update and selective retrieval of data.

A useful way to visualize how data are stored in WIS is to imagine a large cube, made up of individual cells (Tindall and Moore, 1992). The three axes of the cube (see Figure 1) represent features (WHERE observations are made), attributes (which record WHAT the observations are) and occasions (WHEN the observations are made). Thus, each cell in the cube records the value of an attribute at a particular feature for a particular point in time. For example, one cell might record the concentration of calcium on 29 June 1995 at 10:15 (GMT) in the River Swale at Catterick.

WIS regards all values as potentially changeable over time, thus enabling it to handle time series data such as river flows. In the literature, many descriptions for data models can be found (e.g. Sussman, 1993; Hadzilacos and Tryfona, 1996; Tang *et al.*, 1996), but WIS is atypical in that it does not distinguish between spatial and non-spatial data types since it treats all data as potentially time variant. Some examples of WIS data types are:

- real;
- integer;
- text;
- arrays;
- spatial data types (e.g. point, line, grid, etc.); and
- relationships between objects.

Hence, the WIS database is able to hold river flow, water quality, invertebrate, land cover, consents to discharge, licence data, the 1:50 K digitized river network and remotely sensed data, all within a single database. As all attributes can form a time-series, it is possible to store in WIS, the coordinates plus the observed data of research vessel cruises and aircraft flight paths.

There is no constraint on the number of features, attributes or occasions that can be recorded in WIS, other than that imposed by the physical limits of the hardware. The cube is infinite in all directions. Thus, some of the key properties of the cube are:

- any attribute may be observed at any feature;
- a feature may have any number of attributes;
- any number of values may be recorded for an attribute over time at a feature; and
- the values may be recorded at fixed or random time intervals, or in any combination.

THE WIS PHYSICAL DATA MODEL

In using WIS, most users are only aware of the WIS cube concept and are unaware of how data are actually held at the physical level. Although WIS does have an application programmer's interface, this is not currently in the public domain. Writers of application programs therefore need to understand how the data are stored. The physical storage of the data is achieved by a set of relational tables which can be grouped into three types.

Data tables. These contain the observed data. Each value is stored in a cell within the logical WIS cube, at coordinates referenced by the Feature, Attribute and Time.

Reference data. Feature type definitions, attribute definitions, permitted values for attributes and system data.

Lists. Lists represent locations in the WIS cube, for example lists of features (WHERE lists), attributes (WHAT lists) and times (WHEN lists), and combined lists, such as WHERE/WHEN lists are also possible. Lists are created as the result of queries and are an alternative to saving the query itself. Once created, lists enable the marked data to be processed further to produce reports, graphs, etc. They can also be referenced by other queries permitting the orderly building of complex selections.

DATA TABLES

The WIS data tables store the observed data values. These data include the information describing the various features themselves, such as site name and location, together with observations made at the features such as river flow, water temperature or chemical concentrations.

A key point in the physical design of WIS is that the data tables all have the same basic layout, which is shown in Table I. The first four columns locate the 'value' within the cube. FID is the internal identifier for the feature; DID is the identifier for the attribute and YMD (year, month, day) and TIME locate the value in time. They are the coordinates of the value with respect to the three axes of the WIS cube (Feature, Attribute and Time). QCODE (the qualifier), MCODE (the method) and VCODE (the validation status code) modify the interpretation or use of the value. Combined with other information, such as who obtained the data, they record the source and lineage of the information. The final column of the tables contains the attribute's 'value'.

Table I. The general form of all WIS data tables

FID	DID	YMD	TIME	QCODE	MCODE	VCODE	'value'

The values are held in a 'structure'. The structure of the attribute determines the columns needed to hold the value. For each structure, there is a separate table in which values of that type are stored. Most structures contain a single item such as a number or text string, but others, such as position, may hold several values, e.g. X, Y and Z.

REFERENCE DATA

Wherever possible, every aspect of the WIS data model has been designed in a generic manner, and it is this that enables the system to handle a wide range of data types. It has been achieved by allowing the users to define the nature of their data to the system via dictionaries. One of the important lessons learned from LOIS has been that the design of the WIS dictionaries also needs to be generic, as it has become apparent that the

LOIS scientists wish to incorporate their own parameters in the definition. At present, a user defining, for example, the attribute 'concentration of mercury', would specify, *inter alia*:

Attribute Code = 'MERC'
Attribute Name = Mercury
Unit = mg/l
Data Type = Real
etc.

From the system's point of view, this is all that is needed for it to be able to receive, store and output information on mercury. However, the act of building a dictionary often highlights areas of ambiguity in terminology and stimulates the introduction of standards. As a result, the dictionaries become not just a means of making the system generic, but also valuable reference documents that users may want to enhance by adding their own information. For example, additional parameters a chemist might wish to include are chemical formulae, molecular weight and chemical abstract numbers. Similarly, it would be convenient for biologists if, say, the dictionaries of plants and animals could also include identification data. Paradoxically, it appears that the best solution to this problem is to simplify the system by treating the definitions of attributes as features and storing them in the cube. Users can then incorporate any information required in an attribute's definition.

ACCESS TO DATA

To create a database system and the organization to populate it achieves little, if there is no matching arrangement for accessing the data. At the start of LOIS, most users requested their data by telephoning or writing to a data centre and received back data on disk or tape. This was not an ideal solution, since it generated large amounts of work, unnecessary delays, multiple copies of the data around the country and did not give users the benefits of the harmonized and integrated database.

During the course of LOIS, access to the internet has become widespread, opening up the possibility of direct user access to the database. Providing such access is being approached in two steps. In the first step, the system creates and maintains its own meta-data catalogue in the form of a series of web pages. These can be browsed by users who then complete and email requests to the data centres.

The second step, which is being developed at present (see Figure 2), takes the form of a Java geographical interface program that can be down-loaded from the internet. This allows users to interrogate the database itself using GIS-style tools, and then retrieve their data selectively.

Client/Server Web Technology

● **Remote database access via the World Wide Web**

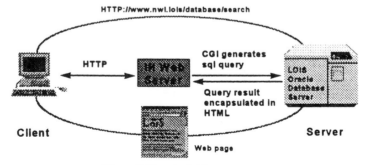

Figure 2. Remote database access

CONCLUSIONS

The objective of LOIS is to understand the processes of the coastal zone in order to enable policy makers and planners to develop sustainable environmental strategies. In a complex and changing world, the planners will inevitably require decision support systems. These will incorporate models that will in turn need to draw on a wide variety of data. For the systems to be viable, the data will need to be harmonized and held in a well-structured database. National and international environmental legislation is ensuring that new and different types of data are required to be monitored. Together, these suggest that databases for storing such data need to be highly generic in nature.

LOIS has provided a unique opportunity to test a generic data model with the use of large volumes of widely varying data types. In general, the model performs well and achieves its objective of being able to handle most environmental data. In terms of speed, relatively rapid data retrieval has been achieved at the expense of slow loading. This is probably reasonable in that most queries are interactive, whereas data loads are batch-mode tasks, probably performed overnight. However, it is now possible to identify additional data types that would be useful, such as arrays and binary large objects. These would be valuable for recording rating tables, borehole lithology, images and documents.

The system was written before the introduction of object oriented (OO) programming. However, most of the WIS database's concepts are entirely compatible with the OO approach. To date, none of the commercially available OO databases have considered the problem of time variant data. It would be interesting to reimplement the WIS cube according to the OO paradigm.

An unplanned but beneficial by-product of using a generic database design is that the process of describing the data to be stored in the system aids the process of standardization. However, the consequence of this is the need to develop dictionary systems that allow the inclusion of user data.

Despite the fact that many data are computerized, the production of meta-data catalogues has so far been a manual task. LOIS has shown that this need no longer be the case and that they can be generated and published directly by the database system itself. However, the need for separate catalogues of data should disappear as users become able to browse databases directly across the internet.

A database, however, is useless without a supporting infrastructure that can respond to users' needs. LOIS has found a workable organizational structure in which one of the keys to success has been delegated funding to the working levels of the data centres, thus enabling quick response to users' needs.

REFERENCES

Browne, T. J, 1995. 'The Role of Geographical Information Systems in Hydrology'. In Foster, I. D. L., Gurnell, A. M., and Webb, B. W. (eds), *Sediment and Water Quality in River Catchments*. John Wiley & Sons Ltd, Chichester. pp. 33–48.

Collet, C., Consuegra, D., and Joerin, F. 1996. 'GIS Needs and GIS Software'. In Singh, V. P. and Fioentino, M. (eds), *Geographical Information Systems in Hydrology*. Kluwer Academic Publishers, Dordrecht. pp. 115–174.

Hadzilacos, T. and Tryfona, N. 1996. 'Logical data modelling for geographical applications', *J. Geog. Inform. Syst.*, **10**, 179–203.

HMSO, 1994. *Sustainable Development: The UK Strategy*. HMSO, London.

Jamieson, D. G. and Fedra, K. 1996, 'The "WaterWare" decision support system for river-basin planning. 1. Conceptual design', *J. Hydrol.*, **177**, 163–175.

Lowry, R. K. and Cramer, R. N. 1995. 'Database applications supporting Community Research Projects in NERC marine sciences', *Geol. Data Manage, Geol. Soc. Spec. Publ.*, **97**, 103–107.

Lowry. R. K. and Loch, S. G. 1995. 'Transfer and SERPLO: powerful data quality control tools developed by the British Oceanographic Data Centre', *Geol. Data Manage., Geol. Soc. Spec. Publ.*, **97**, 109–115.

Moore, R. V. and Tindall, C. I. 1992. 'What is WIS?' *IH/ICL Report*. Institute of Hydrology, Wallingford, UK.

Natural Environmental Research Council (NERC), 1992, *Land-Ocean Interaction Study Implementation Plan*. NERC, Swindon.

Natural Environmental Research Council (NERC), 1996. *NERC Data Policy Handbook Version 1.0 January 1996*. NERC, Swindon.

Sussman, R. 1993. 'Municipal GIS and the enterprise model', *J. Geog. Inform. Syst.*, **7**, 367–377.

Tang, A. Y., Adams, T. M., and Usery, E. L. 1996. 'A spatial data model design for feature-based geographical information systems', *J. Geog. Inform. Syst.*, **10**, 643–659.

3

THE TREATMENT OF FLAT AREAS AND DEPRESSIONS IN AUTOMATED DRAINAGE ANALYSIS OF RASTER DIGITAL ELEVATION MODELS

LAWRENCE W. MARTZ[1]* AND JURGEN GARBRECHT[2]

[1]*Department of Geography, University of Saskatchewan, 9 Campus Drive, Saskatoon, Saskatchewan S7N 5A5, Canada*
[2]*USDA–ARS Grazinglands Research Laboratory, 7207 West Cheyenne Street, El Reno, OK 73036, USA*

ABSTRACT

Methods developed to process raster digital elevation models (DEM) automatically in order to delineate and measure the properties of drainage networks and drainage basins are being recognized as potentially valuable tools for the topographic parameterization of hydrological models. All of these methods ultimately rely on some form of overland flow simulation to define drainage courses and catchment areas and, therefore, have difficulty dealing with closed depressions and flat areas on digital land surface models. Some fundamental assumptions about the nature of these problem topographic features in DEM are implicit in the various techniques developed to deal with them in automated drainage analysis. The principal assumptions are: (1) that closed depressions and flat areas are spurious features that arise from data errors and limitations of DEM resolution; (2) that flow directions across flat areas are determined solely by adjacent cells of lower elevation; and (3) that closed depressions are caused exclusively by the underestimation of DEM elevations. It is argued that while the first of these assumptions is reasonable, given the quality of DEMs generally available for hydrological analysis, the others are not. Rather it seems more likely that depressions are caused by both under- and overestimation errors and that flow directions across flat areas are determined by the distribution of both higher and lower elevations surrounding flat areas. Two new algorithms are introduced that are based on more reasonable assumptions about the nature of flat areas and depressions, and produce more realistic results in application. These algorithms allow breaching of depression outlets and consider the distribution of both higher and lower elevations in assigning flow directions on flat areas. The results of applying these algorithms to some real and hypothetical landscapes are presented.

KEY WORDS   digital elevation models; automated drainage analysis; closed depressions; flat areas; hydrology; watershed

INTRODUCTION

A variety of methods have been developed to process raster digital elevation models (DEM) automatically in order to delineate and measure the properties of drainage networks and drainage basins (Mark 1983; Band, 1986; Jenson and Domingue, 1988; Tarboton *et al.*, 1991; Martz and Garbrecht 1992; Wolock and McCabe, 1995). These methods offer the advantages of speed, precision, reproducibility and the generation of digital output data that are readily incorporated into geographical information systems (Tribe, 1992). They have become recognized as a potentially valuable tool for the topographic parameterization of hydrological models, especially for larger watersheds where the manual measurement of network and basin properties is a time-consuming and error-prone activity (Garbrecht and Martz, 1996).

---

* Correspondence to: Lawrence W. Martz, Department of Geography, University of Saskatchewan, 9 Campus Drive, Saskatoon, Saskatchewan S7N 5A5, Canada.

All methods of raster DEM processing for watershed segmentation and parameterization ultimately rely on some form of overland flow simulation to define drainage courses and catchment areas (Martz and Garbrecht, 1992). Consequently, these methods have difficulty dealing with closed depressions and flat areas on digital models of the land surface. These features contain cells that are completely surrounded by other cells at the same or higher elevation and act as sinks to overland flow. While closed depressions and flat areas in a DEM may represent real landscape features, they are more often artefacts that arise from input data errors, interpolation procedures and the limited horizontal and vertical resolution of the DEM (Mark, 1983, 1988; Jenson and Domingue, 1988; Tribe 1992; Martz and Garbrecht, 1992; Zhang and Montgomery, 1994). Whatever their origin, they require special treatment to allow the complete definition of overland flow patterns across a DEM surface.

A variety of techniques has been proposed to treat closed depressions and flat areas during raster DEM processing for drainage analysis. This paper reviews these techniques and examines the implicit assumptions on which they are based. This leads to the proposal of new algorithms for the treatment of closed depressions and flat areas based on more realistic assumptions about the nature of these features as they are represented in raster DEMs.

FLOW ROUTING

Raster DEM processing for watershed segmentation and parameterization relies fundamentally on the routing of overland flow between the grid cells on which the DEM is structured. This requires that the direction of overland flow between adjacent cells be determined. This can be done in several ways. The most widely used approach, termed the deterministic eight neighbour (D8) method by Fairchild and Leymarie (1991), compares the elevation of each DEM cell with the elevations of its eight adjacent cells (i.e. those within one row and one column). A flow direction is then assigned to point towards one of the eight adjacent cells to which the steepest downslope path exists. Ambiguities that arise when the steepest downslope path is found at more than one adjacent cell are resolved by a variety of secondary decision rules. This method has been criticized on the grounds that the overland flow generated on a raster cell is likely to be distributed between several neighbouring cells in some cases, and several alternative multiple flow direction methods have been proposed (Freeman, 1991; Costa-Cabral and Burges, 1994; Tarboton, 1997).

Regardless of whether the D8 method or one of the several multiple flow direction methods is employed to determine flow direction, closed depressions and flat areas in a DEM pose a problem. The essential difficulty is that these features can contain cells that act as sinks to overland flow because they are completely surrounded by other cells at the same or higher elevation and, therefore, have no downslope path available to any adjacent cell. This presents obvious difficulties for simple, cell-to-cell overland flow routing.

In one sense, there is no meaningful difference between a sink in a closed depression and one on a flat area; both lack topographic information to direct further overland flow routing. However, directing flow from a sink in a closed depression requires somewhat different treatment than is the case for flat areas. Flow proceeding from a sink on a flat area only needs to cross other cells of the same elevation as the sink until a cell with a downslope path is encountered and flow routing can proceed. However, for a sink in a closed depression, flow must cross other cells of higher elevation than the sink to reach a cell with an available downslope path. It should also be noted that closed depressions can be complex features that might contain flat areas (on the floor of the depression or as terraces along the inner slope of the depression), may consist of several local elevation minima separated by ridges and can be surrounded by flat areas.

IMPLICIT ASSUMPTIONS

A variety of methods have been proposed to deal with the problems presented by these topographic features. Each of these makes implicit assumptions about the true nature of the problem topographic features encountered in DEMs. This section attempts to identify these assumptions and examine their viability.

Causes of depressions and flat areas in DEM

Methods for dealing with problem topographic features fall into two broad groups distinguished by their underlying assumption as to the nature of the flat areas and depressions represented in the DEM (Freeman, 1991). One group assumes that flat areas and depressions in a DEM are real landscape features that need to be handled in a hydrologically meaningful way during drainage analysis, while the other views them as spurious features that should be corrected or removed prior to drainage analysis.

The method of Martz and deJong (1988) is representative of the first group. It uses an overland flow simulation technique in which depressions are handled as ponds that fill with water and then overflow. The group of grid cells representing a depression is effectively treated as an individual feature for which a single catchment area value is derived and from which flow proceeds from a single outlet. Flat areas are simply handled as ponds of infinitely small depth.

Freeman (1991) suggests that this approach is appropriate for processing accurate, high resolution and high precision DEMs where the underlying assumption regarding surface features is defensible and where flat areas are of limited extent. Such DEMs are not widely available and can normally be generated only for relatively small areas where detailed topographic measurements are available. As a result, application of this method has been largely restricted to field-scale erosion studies (Martz and deJong, 1991). This type of approach has the potential, as yet unrealized, to consider the relationship between the inflow volume and storage capacity of surface depressions and, thereby, to incorporate the impact of depression storage on derived drainage patterns.

The second group of methods deals with flat areas and depressions under the assumption that they are spurious and should be corrected prior to analysis. This assumption seems reasonable for most readily available DEMs. These are often based on elevation data derived from conventional topographic maps and generated using relatively simple interpolation methods. Spurious features can arise because of low quality input data, interpolation errors during DEM generation, truncation or rounding of interpolated values to lower precision and averaging of elevation information within grid cells (Martz and Garbrecht, 1992; Tribe, 1992).

Several different treatment methods fall into the group that assumes flat areas and depressions are spurious DEM features. The simplest method uses a smoothing filter to remove as many problematic features as possible, and, in some cases, arbitrarily removes or adjusts any that survive the smoothing process (Mark, 1983). The primary difficulty with this approach is that it reduces the overall information content of the digital elevation data by uniformly applying corrective measures to both problematic and non-problematic areas of the DEM (Tribe, 1992).

Methods that focus corrective measures solely on the problem features are more widely used. These typically follow a two-step procedure similar to that introduced by Jenson and Domingue (1988). The DEM is first modified by raising the elevations of the cells in each closed depression to the elevation of its lowest outlet (i.e. a cell where a downslope path is available to a cell outside the depression). This simulates filling all depressions to replace them with flat areas. Flow directions are then assigned to the cells in all flat areas, both those produced by depression filling and those present in the original DEM, so as to direct flow to an outlet on the edge of the flat area.

Flow directions

The most widely used approaches to assigning flow directions in flat areas follow that introduced by Jenson and Domingue (1988). Flat area cells are scanned to find those adjacent to other cells with a defined flow direction. These flat area cells are then assigned a flow direction pointing to the nearest adjacent cell with a defined flow direction. This is repeated until all flat area cells are assigned a flow direction. The relief imposition method of Martz and Garbrecht (1992) for deriving flow directions gives equivalent results. Both of these approaches constrain flow paths to remain within the flat area and allow the possibility of multiple outlets. However, as Tribe (1992) points out, they often produce unrealistic, parallel flow patterns over flat areas.

Tribe (1992) suggests an alternative method of flow direction assignment to enhance flow convergence. This method essentially involves defining a main flow path through the flat area and directing other flow paths towards this main path. Unfortunately, because the main flow path is defined along the shortest path between the inflow and the nearest outlet, it is possible for the main flow path to pass through areas of higher elevation to reach the outlet. This appears to introduce new problems as much as it resolves existing ones (Martz and Garbrecht, 1995).

Both the approach of Jenson and Domingue (1988) and that of Tribe (1992) can produce unrealistic flow patterns on flat areas, and, while the shortcomings of these two approaches are somewhat different, they arise from a common, implicit assumption. This is that flow directions across a flat area are determined solely by adjacent cells of lower elevation. But if we accept that flat areas are spurious features for which topographic information is incomplete, it seems more reasonable to expect flow to be directed both towards lower elevation and away from higher elevation.

Depression-filling

It is also important to recognize that the depression filling that produces many of the flat areas over which flow direction assignment is required makes several implicit assumptions about the cause of DEM depressions. If all closed depressions in a DEM are assumed to be spurious, they must result from either over-estimating or underestimating the elevation values for some DEM cells. The specific effect of both underestimation and overestimation errors on flow patterns is dependent on the topographic settings in which they occur. In general, however, underestimates are expected to produce indentations in the surface, while overestimates are expected to produce blockages or dam-like features along flow paths (Figure 1).

Even though closed depressions result from both underestimation and overestimation errors, the practice of eliminating these features solely by simulated filling to the level of the local outlet implicitly assumes that all depressions arise from underestimation errors. This would seem to introduce systematic bias into the modified DEM. Ideally, closed depressions caused by underestimation errors should be eliminated by filling, while those caused by overestimation errors should be eliminated by outlet lowering. However, implementation of such treatment faces the problem that it is impossible to distinguish between underestimation and overestimation errors directly by evaluation of a raster DEM.

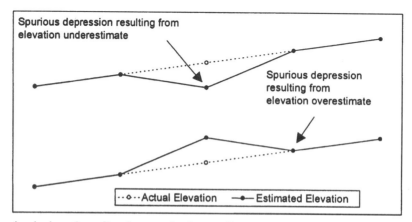

Figure 1. Two-dimensional schematic profiles along a valley bottom illustrating spurious depressions arising from elevation value underestimates and from elevation value overestimates

Summary

Of the following three major assumptions that underlie present methods for handling closed depressions and flat areas in DEM processing for drainage analysis, one appears to be reasonable but the others do not. The assumption that all depressions and flat areas are spurious features underlies all methods except that of Martz and deJong (1988). Most readily available DEMs are subject to various types of error and the representation of landscape features in a DEM is limited by the horizontal resolution and the precision to which elevation values are stored. Given these considerations and the fact that it is not possible to distinguish between spurious and non-spurious features in a DEM, this seems a conservative, but reasonable assumption for processing all but the most detailed and accurate DEMs.

The second major assumption is that flow across flat areas will be directed solely towards lower elevations. This seems less reasonable, in large part because assigning flow directions under this assumption can produce unrealistic flow patterns. Furthermore, it seems that this assumption gives inadequate consideration to the nature of the topography surrounding flat areas. The third major assumption is that all depressions arise from underestimation errors. However, it is clear that depressions can arise from elevation over-estimates as well as from elevation underestimates.

ALTERNATIVE APPROACHES

This paper presents two new algorithms for handling closed depressions and flat areas in DEM processing for drainage analysis. These algorithms operate under the assumption that all depressions and flat areas are spurious, but make different and more realistic assumptions than earlier methods about the nature of flow paths on flat areas and the nature of the errors that produce surface depressions. The first algorithm simulates breaching of the outlet of closed depressions to eliminate or reduce those expected to have been produced by elevation overestimates. The second algorithm modifies flat surfaces to produce more realistic and topographically consistent drainage patterns than those provided under the assumption that flow will move along the shortest available path to the outlet.

While these algorithms could be used to improve any raster-based drainage analysis approach, this paper focuses on their application with a D8 flow routing approach. This focus arises, in part, because the algorithms were developed for incorporation into the DEM processing software module DEDNM (Martz and Garbrecht 1992) which is based on a D8 approach and is a core component of the TOPAZ landscape analysis tool. TOPAZ (topographic parameterization) is designed for topographic evaluation, drainage identification, watershed segmentation and subcatchment parameterization (Garbrecht and Martz, 1995). The more important reason for this focus, however, is that the D8 approach is the most widely used for automated drainage analysis (Tribe 1992) and, despite its limitations (Costa-Cabral and Burges, 1994), provides a reasonable representation of flow patterns for convergent flow conditions (Freeman, 1991) and maintains consistency between flow patterns, calculated contributing area and spatial representation of subcatchments (Martz and Garbrecht, 1992).

THE BREACHING ALGORITHM

The new breaching algorithm is incorporated into the initial preprocessing stage of the TOPAZ software system during which all closed depressions in the DEM are replaced by flat areas. It evaluates the local outlet of each closed depression in a DEM to determine whether the elevation of one or two cells at the outlet could be lowered to eliminate or reduce the size of the depression without reversing the direction of overland flow across the outlet. This effectively simulates breaching of the outlet. Breaching is constrained to one or two cells in the expectation that major flow path obstructions caused by elevation overestimates are likely to be of limited spatial extent. This constraint also prevents long, straight channel reaches being artificially imposed during the breaching process. As a result, the effect of breaching on the overall flow patterns indicated by the

DEM data is minimized. The algorithm does not directly distinguish between overestimation and under-estimation errors in a DEM, but simply uses elevation lowering to eliminate or reduce closed depressions that can reasonably be expected to have resulted from overestimation errors.

Identification of potential breaching sites

The term sink is used to refer to any DEM cell for which no flow direction can be assigned under the D8 method. The term inflow sink is used to refer to a specific type of sink; a sink with at least one adjacent cell at a higher elevation. In other words, inflow sinks are cells that receive inflow but have no outflow. Inflow sinks are found within closed depressions and on the edge of flat areas that receive inflow from cells at higher elevation. The process of identifying potential breaching begins by scanning the DEM to locate inflow sinks. Searching only for inflow sinks, rather than for all sinks, is done for the purpose of computational efficiency. It precludes the subsequent, non-productive analysis of flat summit areas.

A contributing area is defined for each inflow sink as it is encountered in the DEM scan. This is the set of contiguous cells from which water could reach the inflow sink by flowing downslope and across level areas. This set of contiguous cells is defined using a procedure that effectively 'grows' the contributing area outwards from the inflow sink as follows. A window (initially 5 by 5 cells) is centred on the inflow sink and the inflow sink cell is flagged. The window is then scanned and all cells that are adjacent to a flagged cell and at the same or higher elevation as the flagged cell are flagged. The scan is repeated until no additional cells can be flagged (Figure 2a).

The cells in this contributing area are then evaluated to determine if they are potential outlets. Potential outlets are those cells within the contributing area of the inflow sink that are: (1) adjacent to a cell outside the contributing area; and (2) at a higher elevation than a cell outside the contributing area. The lowest of these potential outlet cells is then identified (Figure 2a). If more than one potential outlet cell shares the lowest elevation, the one with the steepest slope out of the contributing area is selected. If no potential outlets are found, then the window does not fully encompass the flat area or closed depression of which the inflow sink is a part. This would also be the case if any contributing area cell along the edge of the window were at a lower elevation than that of the cell identified as the lowest potential outlet. If either of these conditions is encountered, the window is enlarged and the entire procedure for identifying the contributing area is repeated. This continues until the window fully encompasses the depression or flat area around the inflow sink.

Breaching procedure

Once the window has been expanded to the appropriate dimensions and the lowest outlet from the contributing area of the inflow sink has been defined, the contributing area is evaluated to determine if it

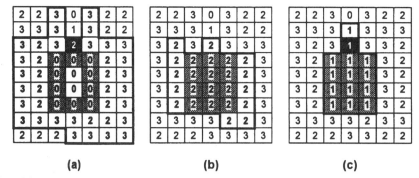

(a)   (b)   (c)

Figure 2. Illustration of depression filling. (a) Original elevations with inflow sink cells (hatched), the lowest potential outlet cell (black) and the sink contributing area (outlined); (b) traditional depression filling result with cells raised by filling (hatched) and flat area (outlined); (c) new breaching and filling result with cells raised by filling (hatched), cells lowered by breaching (black) and flat area (outlined)

contains at least one cell at a lower elevation than the lowest outlet. This would indicate that the contributing area contains a closed depression. If the contributing area does not contain a closed depression, it is a flat area and further analysis to determine the potential for outlet breaching is not appropriate. Cells within the contributing area and at the elevation of the lowest outlet are simply flagged to prevent subsequent, redundant evaluations being initiated at any other inflow sinks on the perimeter of the flat area.

The breaching procedure is initiated only if the contributing area of an inflow sink is found to contain a closed depression. This prevents the simulated incision of a drainage course into a flat area. The simulated incision of a drainage course is permitted only through an area of higher elevation separating two areas of lower elevation, as would be the case if the water impounded behind a topographic obstruction were to overflow and erode through the obstruction, causing a breach. Such simulated breaching of topographic obstructions encountered along a flow path does not have the effect of substantially changing drainage divides, as the breach occurs along the same path as would be followed by water spilling from the closed depression. The number of DEM cells that may be lowered to simulate breaching is termed the breaching length. To restrict breaching to relatively narrow obstructions, the breaching length is limited to a maximum value of two cells.

The actual breaching occurs as follows. All cells that are inside the contributing area of the inflow sink and at the same elevation as the lowest outlet are examined to determine if they are adjacent to a cell that is both outside the contributing area and at a lower elevation than the outlet. Any overland flow originating on or entering such cells would proceed to a cell outside the contributing area and, therefore, away from any closed depression it contained. These cells are then examined to find if they are within the breaching length of a cell which is inside the contributing area and at a lower elevation than the lowest outlet. This identifies cells at which a drainage course no longer than the breaching length could be incised through the topographic barrier around the closed depression. All cells meeting both of these criteria are potential breaching sites. If no such cells exist, no breaching is possible and all closed depressions within the contributing area are simply filled to the elevation of the lowest outlet. If more than one potential breaching site exists, the one with the steepest slope to a cell outside the contributing area (primary criterion) and the steepest slope to a cell inside the contributing area (secondary criterion) is selected. If these criteria are met at more than one site, the one encountered first is selected arbitrarily.

The elevation of the cell at the selected breaching site is lowered to the elevation of the cell outside the contributing area to which it flows. For a two-cell breaching length, the elevation of the next cell along the most direct path to the inside of the contributing area is also lowered to this elevation. This changes the elevation of the outlet and effectively breaches the obstruction responsible for the closed depression (Figure 2c). Lowering the elevation of the cells along the breaching path to the elevation of the cell outside the contributing area precludes the possibility of flow entering a closed depression through what would otherwise have been the outlet. In cases where flow from the outlet is towards a cell undefined for elevation, breaching is still permitted, but the elevation of the cell(s) along the breaching path is set to the elevation of the nearest cell inside the contributing area.

Depression filling

Closed depressions may still exist within the contributing area of an inflow sink after a breach is effected. Closed depressions for which a breach is effected might be simply reduced in spatial extent or even remain unchanged in spatial extent (Figure 2c). Therefore, the next step in the analysis of a contributing area containing one or more closed depressions is done regardless of whether a breach has been effected. This involves changing the elevation of all cells that are both inside the contributing area and lower than the selected outlet, to the elevation of the outlet. Where a breach has been effected, the breaching site will now be the outlet. This effectively fills all closed depressions in the contributing area to produce a continuous flat surface at the elevation of the outlet (Figure 2b). The cells in this flat area are flagged to prevent subsequent, redundant evaluations being initiated at other inflow sinks on the perimeter of the flat area produced by depression filling.

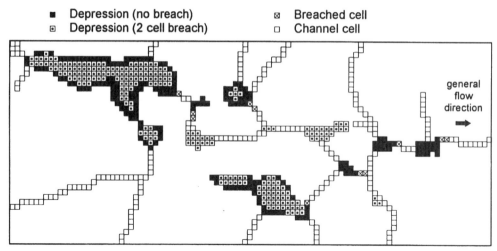

Figure 3. Illustration of the application of the depression outlet breaching algorithm to a typical, low relief landscape in central Oklahoma

Example application

The new breaching algorithm was extensively tested during development in a variety of real and hypothetical topographic settings and found to be robust and to provide satisfactory results, even when dealing with complex, nested and truncated depressions. Figure 3 illustrates the result of the application of the breaching algorithm to an area of typical, low relief topography in central Oklahoma. The effect of breaching was both the complete removal of some smaller depressions and a decrease in the size of the larger depressions. Some depressions were unaffected since no breaching of these features is possible under the breaching length constraint. A visual assessment of this illustration suggests that the effect of breaching is intuitively satisfactory and that no unusual consequences result. In the context of the broader purpose of the delineation and parameterization of drainage systems, breaching reduced the size of the depressions that would be filled to produce flat areas on which drainage directions need to be imposed.

THE FLAT AREA ALGORITHM

The new flat area algorithm is also incorporated into the initial preprocessing stage of the TOPAZ software system. Here, we review the major features of the algorithm in the context of the implicit assumptions underlying automated drainage analysis of raster DEMs. A more detailed description of the algorithm is available in Garbrecht and Martz (1997).

The flat area algorithm modifies cell elevations in flat areas to produce desired drainage directions during the subsequent DEM processing. Elevation modifications are based on the assumption that drainage directions are oriented both towards lower elevation and away from higher elevation. Such a drainage pattern is obtained by imposing two gradients on the flat surface; one towards adjacent areas of lower elevation, which draws flow to the nearest outlet, and a second that forces flow away from adjacent areas of higher elevation. For flat areas predominantly surrounded by higher terrain, this produces a convergent drainage pattern. However, the slope steepness of the surrounding higher and lower terrain does not influence the imposed gradients.

Elevation increment

The elevation modifications are made by the successive addition of an elevation increment to the initial cell values. The increment is very small relative to cell elevation precision and, while sufficient to identify

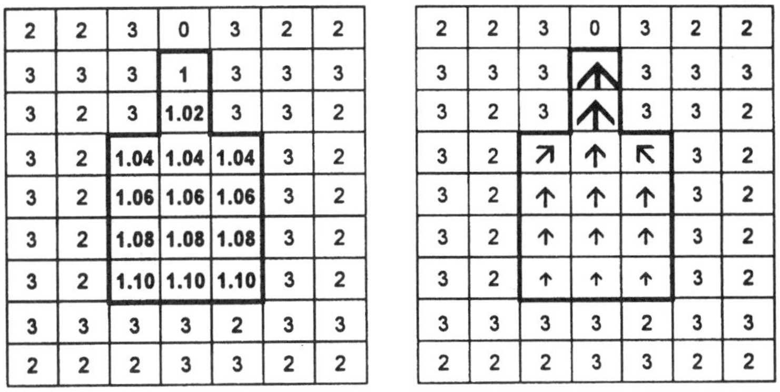

Figure 4. Illustration of imposition of gradient on flat area towards lower elevations using an elevation increment of 0·02. (a) Addition of cumulative elevation increments and initial elevations on outlined flat area; (b) resulting flow directions with arrow size proportional to upstream drainage area

flow direction over the flat surface, does not significantly alter the elevation values in the digital landscape model.

The elevation increment used is 0·00002 units of the DEM elevation values. Thus, for a DEM with elevations in units of metres, one elevation increment is 20 μm. The addition of such a small elevation increment produces a relatively minor change in actual cell elevation, yet is adequate to define a flow direction numerically. The basic increment value was set at 0·00002 units rather than 0·00001 units, because the latter value is needed for the treatment of exceptional situations, that are discussed later.

In practice, all elevation values are handled internally as 32-bit integers to avoid any possible rounding errors associated with using such small elevation increments. Input DEM values are required to be in units of metres and limited to the range 1·0 to 9999·0. They are read to a precision of 1 decimetre and multiplied by 100 000. Therefore, an input elevation value of 17·43 m would be stored internally as 1 740 000 and the basic elevation increment would be 2.

Gradient towards lower terrain

The gradient towards lower terrain is imposed by incrementing the elevation of all cells on the flat surface that are not adjacent to a cell with a lower elevation or an existing downslope gradient. The increment is applied again to all cells that still have no downslope gradient. This is repeated until all cells have a downstream gradient (Figure 4a). Note that, in the interests of clarity, the elevation increment used in this and the two subsequent figures is 0·02 elevation units rather than 0·00002 elevation units, as employed in the algorithm. In this way, a flow gradient towards lower terrain is constructed as a backward growth from the outlet(s) into the flat area. The drainage resulting from the imposed gradient is shown in Figure 4b. This is equivalent to the final drainage pattern that would be derived using the Jenson and Domingue (1988) approach to drainage direction assignment in flat areas, and illustrates the unrealistic parallel flow pattern discussed by Tribe (1992).

Gradient away from higher terrain

The gradient away from higher terrain is imposed by first incrementing the elevation of all cells in the flat surface that are adjacent to higher terrain and have no adjacent cell at a lower elevation. The imposed increment introduces a downslope gradient away from higher terrain for all cells immediately adjacent to higher terrain. In subsequent passes the increment is applied as follows: (1) all cells that have been incremented previously are incremented again; and (2) all cells that have not been incremented previously and that are adjacent to an incremented cell and that are not adjacent to a cell lower than the flat surface, are

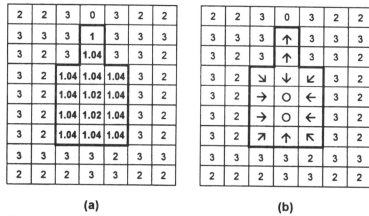

(a)   (b)

Figure 5. Illustration of imposition of gradient on flat area away from higher elevations using an elevation increment of 0·02. (a) Addition of cumulative elevation increments and initial elevations on outlined flat area; (b) resulting flow directions as arrows and sinks as circles

incremented. This step is repeated until all cells in the flat surface have been incremented (Figure 5a). This produces a gradient away from higher terrain by inward growth from the edges of the flat surface that are adjacent to higher terrain. The drainage pattern corresponding to this gradient is shown in Figure 5b. Because the flat surface is surrounded by higher elevations most flow directions point away from higher terrain. Exceptions are those cells adjacent to non-incremented cells.

Combined gradient and final drainage pattern

In a third and last step, the increments applied in the previous two steps are added together for each cell to define a total increment. This total increment is then added to the initial elevation of each cell producing a surface that is no longer flat and includes: (1) a gradient away from higher terrain; and (2) a gradient towards lower terrain. One additional adjustment may be necessary in some cases. The imposed gradients towards lower terrain and away from higher terrain can be in exactly opposite directions and of the same magnitude. They cancel each other, leaving a cell without a downslope path to a neighbouring cell. For these rather exceptional situations an additional one-half basic increment (i.e. 0·00001 elevation units) is added to the problem cell. Using a full increment is not practical because it could, under some circumstances, create another flat surface with upstream cells that are exactly one increment higher than the problem cell. If this situation were to encompass several adjacent cells, then the one-half increment is added repeatedly following the same procedure as is used in imposing the gradient towards lower terrain.

Figure 6a illustrates the result of combining the gradients imposed in the first two steps of the analysis. In this example, the cell at row 5, column 4 proved to be a problem cell with an initial modified elevation of 1·08. Since this would not allow a downslope path to an adjacent cell, one-half increment was added to produce an elevation value of 1·09 and provide a downslope gradient. The resulting flow pattern is shown in Figure 6b. This differs from the flow pattern in Figure 4b in that flow tends to converge and concentrate along the central axis of the flat area.

Example application

The change in drainage patterns between Figure 4b (the traditional approach) and Figure 6b (the new approach) is not very pronounced because of the small size of the example data set. A more dramatic illustration of the effect of the new algorithm on drainage patterns is given in Figures 7 and 8. These show a saddle consisting of a flat surface between higher terrain to the right and left, and three locations of lower terrain. Additional complications are introduced by a wedge of higher terrain protruding into the flat surface from the bottom, and a rectangular indentation of the flat surface into higher terrain at the top right corner

2	2	3	0	3	2	2
3	3	3	1	3	3	3
3	2	3	1.06	3	3	2
3	2	1.08	1.08	1.08	3	2
3	2	1.10	1.09	1.10	3	2
3	2	1.12	1.10	1.12	3	2
3	2	1.14	1.14	1.14	3	2
3	3	3	3	2	3	3
2	2	2	3	3	2	2

(a)

2	2	3	0	3	2	2
3	3	3	↑	3	3	3
3	2	3	↑	3	3	2
3	2	↗	↑	↖	3	2
3	2	↑	↑	↑	3	2
3	2	↗	↑	↖	3	2
3	2	↗	↑	↖	3	2
3	3	3	3	2	3	3
2	2	2	3	3	2	2

(b)

Figure 6. Illustration of combination of gradients away from higher elevations and towards lower elevations using an elevation increment of 0·02. (a) Addition of cumulative elevation increments for both gradients and initial elevations on outlined flat area and adjusted elevation (underlined) for special case cell; (b) resulting flow directions as arrows proportional to upstream drainage area

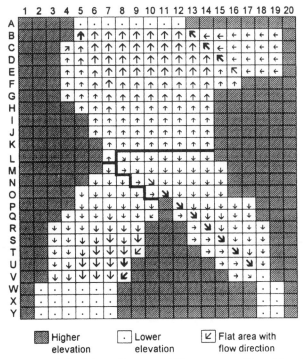

Higher elevation   · Lower elevation   ↙ Flat area with flow direction

Figure 7. Illustration of the flow pattern derived for a flat area within a topographic saddle using the method of Jenson and Domingue (1988)

of the figure. The drainage pattern imposed by the method of Jenson and Domingue (1988) is shown in Figure 7 and that imposed by the new algorithm is shown in Figure 8. The drainage pattern imposed on the initial flat surface is indicated by arrows, which are proportional in size to the upstream drainage area.

The drainage in Figure 7 shows a parallel flow pattern. Drainage paths running towards the top outlet tend to be parallel with little indication of convergence. Flow from higher elevation on to the flat surface tends to follow immediately along the edge of the higher area so that sites with the largest upstream drainage area are not in the middle of the saddle, but at the right and left edge of the top outlet. The bottom-right portion of

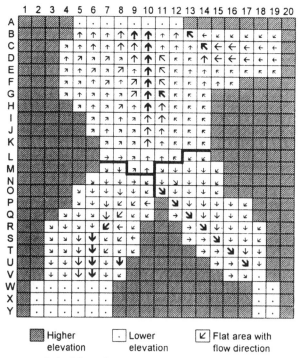

Figure 8. Illustration of the flow pattern derived for a flat area within a topographic saddle using the proposed new algorithm for drainage direction assignment

the saddle has a major drainage path running at a 45° angle with parallel flow moving toward it. The 45° main drainage path is directed to the outlet cell in the lower-right corner and is not influenced by higher terrain surrounding the flat surface.

The drainage pattern in Figure 8 is towards all three locations of lower terrain and away from higher terrain. A drainage divide is situated near the middle of the saddle approximately midway between the locations of lower terrain. Flow convergence is well developed and consistently away from higher elevations and towards lower elevations as one would intuitively expect in this topographic situation.

CONCLUSIONS

Most existing methods for handling closed depressions and flat areas in DEM processing for drainage analysis are based on some common and fundamental assumptions about the nature of these features. These assumptions are largely implicit to the methods and are: (1) closed depressions and flat areas are spurious features that arise from data errors and limitations of DEM resolution; (2) flow directions across flat areas are determined solely by adjacent cells of lower elevation; and (3) closed depressions are caused exclusively by the underestimation of DEM elevations. While the first of these seems reasonable, the others do not.

It is possible to make more reasonable assumptions about the controls on flow direction and the cause of closed depressions, and to incorporate these assumptions into new algorithms for handling difficult topographic situations encountered in raster DEM processing for drainage analysis. A method that combines depression breaching and filling is proposed to remove spurious sinks in a DEM. Although the method does not explicitly distinguish depressions resulting from elevation overestimation from those arising from elevation underestimation, it uses breaching to eliminate or reduce depressions that can reasonably be expected to have resulted from elevation overestimation. The method proposed to define drainage over flat areas in a DEM uses information on the surrounding topography and allows flow convergence within the flat area.

The two new algorithms presented here are based on a deductive, but qualitative, assessment of the most probable nature of depressions and flat areas in raster DEMs. In application, the algorithms have proved to be robust and able to handle all types of actual and hypothetical topographic configurations encountered to date. The algorithms also provide results that are intuitively more satisfactory and more realistic than those of earlier methods.

REFERENCES

Band, L. E. 1986. 'Topographic partition of watersheds with digital elevation models', *Wat. Resourc. Res.*, **22**, 15–24.

Costa-Cabral, M. C. and Burges, S. J. 1994. 'Digital elevation model networks (DEMON): a model of flow over hillslopes for computations of contributing and dispersal areas', *Wat. Resourc. Res.*, **30**, 1681–1692.

Fairchild, J. and Leymarie, P. 1991. 'Drainage networks from grid digital elevation models', *Wat. Resourc. Res.*, **27**, 29–61.

Freeman, T. G. 1991. 'Calculating catchment area with divergent flow based on a regular grid', *Comp. Geosci.*, **17**, 413–422.

Garbrecht, J. and Martz, L. W. 1995. 'TOPAZ: an automated digital landscape analysis tool for topographic evaluation, drainage identification, watershed segmentation and subcatchment parameterization', *TOPAZ User Manual*, USDA–ARS Publication No. NAWQL 95-3. USDA–ARS, Durant, OK. 110 pp.

Garbrecht, J. and Martz, L.W. 1996. 'Digital landscape parameterization for hydrologic applications', in: Kovar, K. and Nachtnebel, H. P. (eds), *Applicatior of Geographic Information Systems in Hydrology and Water Resources Management, IAHS Publ.*, **235**, 169–174.

Garbrecht, J. and Martz, L. W. 1997. 'The assignment of drainage direction over flat surfaces in raster digital elevation models', *J. Hydrol.*, **193**, 204–213.

Jenson, S. K. and J. O. Domingue, 1988. 'Extracting topographic structure from digital elevation data for geographical information system analysis', *Photogramm. Engng Remote Sens.*, **54**, 1593–1600.

Mark, D. M. 1983. 'Automatic detection of drainage networks from digital elevation models', *Cartographica*, **21**, 168–178.

Mark, D. M. 1988. 'Network models in geomorphology', in: Anderson, M. G. (ed.), *Modeling Geomorphological Systems*. John Wiley & Sons, New York. pp. 73–96.

Martz, L. W. and deJong, E. 1988. CATCH: a Fortran program for measuring catchment area from digital elevation models. *Comp. Geosci.*, **14**, 627–640.

Martz, L. W. and deJong, E. 1991. 'Using cesium-137 and landform classification to develop a net soil erosion budget for a small Canadian prairie watershed', *Catena*, **18**, 289–308.

Martz, L. W. and Garbrecht, J. 1992. 'Numerical definition of drainage network and subcatchment areas from digital elevation models', *Comp. Geosci.*, **18**, 747–761.

Martz, L. W. and Garbrecht, J. 1995. 'Automated recognition of valley lines and drainage networks from grid digital elevation models: a review and a new method — comment. *J. Hydrol.*, **167**, 393–396.

Tarboton, D. G. 1997. 'A new method for the determination of flow directions and contributing areas in grid digital elevation models', *Wat. Resourc. Res.*, **33**, 309–319.

Tarboton, D. G., Bras, R. L., and Rodriquez-Iturbe, I. 1991. 'On the extraction of channel networks from digital elevation data. *Hydrol. Process.*, **5**, 81–100.

Tribe, A. 1992. 'Automated recognition of valley lines and drainage networks from grid digital elevation models: a review and a new method', *J. Hydrol.*, **139**, 263–293.

Wolock, D. M. and McCabe, G. J. 1995. 'Comparison of single and multiple flow direction algorithms for computing topographic parameters in TOPMODEL', *Wat. Resourc. Res.* **31**, 1315–1324.

Zhang, W. and Montgomery, D. 1994. 'Digital elevation model grid size, landscape representation and hydrologic simulations', *Wat. Resourc. Res.*, **30**, 1019–1028.

4

A PHENOMENON-BASED APPROACH TO UPSLOPE CONTRIBUTING AREA AND DEPRESSIONS IN DEMs

WOLFGANG RIEGER*

Institute of Surveying, Remote Sensing, and Land Information, University of Agricultural Sciences, Vienna, Austria

ABSTRACT

Description of the terrain surface through digital elevation models (DEMs) strongly depends on data collection methods and DEM data structures. For efficiency and availability reasons regular point distributions are most common, which yield artefacts such as depressions and preferential flow directions. These facts need to be considered when natural phenomena are modelled, as is shown for handling depressions and for estimation of flow paths and upslope contributing areas. Analysis of the main reasons for the occurrence of depressions shows that they usually better reflect the terrain than their surroundings. Thus, the most common remedial method of raising depressions is rejected. Algorithms that 'cut' a flow path from the depression through its bounding barrier are favoured instead. Several flow routing algorithms are evaluated for their behaviour in regular grids. It is shown that the multiple flow direction (mfd) algorithm that distributes water from a grid cell to the lower of its eight neighbours proportionally to their elevation differences (slope) exhibits correct flow directions and the best rotation invariance. It is suggested that the estimation of upslope contributing areas (TCAs) is undertaken in two steps: first, a high quality flow direction data set is derived by a well-behaved mfd algorithm or by subgrid modelling of flow paths; secondly, the upslope contributing areas are obtained by counting the upslope elements.

KEY WORDS   DEM; upslope contributing area; depressions; flow routing

INTRODUCTION

It is nearly three decades since Freeze and Harlan (1969) outlined a blueprint for physically based hydrological response models. Their ideas have been partly implemented in the following decades, especially with the development of GIS (geographical information systems). However, GIS are mainly used to organize and visualize input and output data; they have not brought significant improvement in modelling strategies and methodology (Wilson, 1996). In particular, several well-known problems have not yet been resolved satisfactorily. Here, aspects of terrain surface modelling with digital elevation models (DEMs) are considered.

Although DEMs are by far the most accurate and dense data available for hydrological response or erosion models, there is still a need for better descriptions of the terrain surface. Several authors have found that a ground resolution of approximately 10 m, depending on the terrain characteristics, is sufficient to describe the variability of the relevant parameters for typical input elevation data (Zhang and Montgomery, 1994; Quinn *et al.*, 1995). Higher resolution might eventually provide variability at even smaller scales (Montgomery, 1996). With conventional data collection methods such high resolution is the absolute limit obtainable at reasonable costs at the catchment scale. For large areas, the use of manual or analytical

---

* Correspondence to: Wolfgang Rieger, Institut für Vermessung, Fernerkundung und Landinformation, Universitaet fuer Bodenkultur, Peter-Jordan-Strasse 82, A-1190, Vienna, Austria. E-mail address: W. Rieger (rieger@edv1.boku.ac.at).

photogrammetry or GPS (Global Positioning System) are too expensive. Digital image matching allows much automation, but has problems with off-terrain objects. All these methods more or less fail in forested areas. A very promising new data collection technology is laser altimetry for both its extremely high resolution (down to submetre) and its potential to penetrate the tree crown layer and obtain accurate elevation data even in densely forested areas (Lohr and Eibert, 1995; Kilian *et al.*, 1996; Kraus *et al.*, 1997; Næsset, 1997). The technique may well become an important elevation data source in the future, especially in forested areas.

Mark (1979) stated, nearly two decades ago, that the phenomena in question should determine model structures and algorithms in use rather than computational considerations. Today, even with high resolution data, it seems that major problems remain. Inappropriate surface descriptions follow not least from arbitrary point distributions during data collection. Grid data structures are the main DEM source in use for historical as well as practical reasons, and it is probable that these inappropriate structures will remain for a long time, especially with high resolution data. So the question is, how to achieve the best results with the existing limitations in data and structure.

It is necessary to address the weaknesses in all stages of the modelling process and to focus upon both the causes of problems and their consequences. Algorithms can be optimized and tuned to improve some of their major drawbacks and reduce quality loss. In the following discussion, these issues are addressed in relation to the handling of depressions and to flow routing with regard to estimating flow paths and upslope contributing area. The main focus here lies on geometric aspects rather than on flow physics. Grid structures are employed because of their frequent use and their persistent and possibly increasing importance in all kinds of terrain modelling.

DEPRESSIONS

General

Depressions in DEMs have been recognized as one of the major problems in hydrological applications, since they hinder flow routing (O'Callaghan and Mark, 1984; Jenson and Domingue, 1988). A depression is defined as a point or an area lower than all immediately surrounding points. Each depression has its own catchment area. The outlet of a depression is defined as a point through which water could leave the depression; normally the lowest point along the border of the depression catchment (e.g. Mark *et al.*, 1984; O'Callaghan and Mark, 1984; Band 1986; Jenson and Domingue, 1988). The area within the depression that is lower than the outlet point is referred to as the 'inner area' of the depression (Figure 1).

The definition of the outlet may be problematic if there are several candidates along the catchment border whose elevations differ only within the height error. Therefore, Jenson and Domingue (1988) proposed an

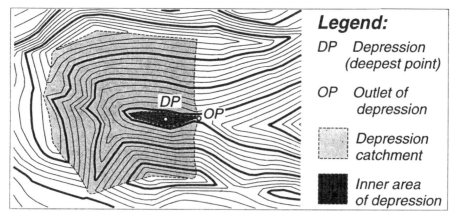

Figure 1. A depression: inner area, outlet and depression catchment

algorithm that allows for several outlets by topologically splitting the area, especially for relatively flat areas. However, within depressions there is no geometric justification for splitting (i.e. from the elevation data). Generally, mere topological treatment of geometrically defined features ignores the broader geometric information that is available. For example, a more regional curvature analysis could be applied to find the most probable border position within depressions or flat areas, but the improvements achieved are probably small compared with the effort. Furthermore, whatever the approach, it is impossible to extract information reliably that is beyond the scope of the data (in terms of scale, resolution and accuracy).

Reasons for depressions in the DEM

Depressions may reflect genuine terrain forms; in such cases water would either fill the depressions and eventually overflow or would find subsurface paths for runoff. Assuming those subsurface paths correspond to the larger terrain structure, these depressions may be eliminated from the DEM in order to define flow paths within the wider area. The depression is thus seen as a local feature that is of minor interest for the delineation of flow paths. Few natural depressions cannot be treated this way, although karst terrain, for example, may provide some exceptions. However, the vast majority of depressions do not reflect genuine terrain features. There are two main reasons for such 'artificial depressions', data errors and the consequences of surface representation.

(a) *Data errors*. The type of errors depends mainly on the data collection method. Typical errors include:

(i) For *manual profiling* (analytical plotters, softcopy photogrammetry), the so-called 'scan error' (Kraus, 1984), resulting in (regularly) shifted profiles, depending on the scan direction. These can be eliminated by appropriate algorithms (Kraus, 1984).

(ii) For *automatic DEM generation* (image matching), off-terrain 'objects', such as vegetation or buildings, tend to raise the surface, frequently resulting in artificial gorges or depressions along areas with ground sight (roads, etc.). Manual correction is necessary in areas where such errors are too high (Walker and Petrie, 1996).

(iii) Random data errors may especially lead to depressions in flat areas where the mean elevation error exceeds the elevation difference between neighbouring points.

(b) *Surface representation owing to point distribution*. The surface representation depends on data collection methods and the DEM data structure used. Ideally, the terrain surface is collected through its main features (skeleton lines). However, this is a time-consuming task that requires high expertise, since the proper location and number of breaklines are neither easy to recognize nor to define and depend widely on the operator's experience (Kubik, 1988). Collecting the terrain surface in (regularly spaced) profiles is much faster and easier. It does not require any information about the surface; thus, automatic DEM measurement techniques (image matching) usually produce grids (Hahn and Förstner, 1988; Helava, 1988; Krzystek, 1991; Ackermann, 1992; Kaiser *et al.*, 1992; Miller and DeVenecia, 1992; Baltsavias *et al.*, 1996), while automatic extraction of skeleton information has only recently been described (Wild and Krzystek, 1996). Hence, improper point distributions and data structures alone, mainly regular grids, are often used.

Owing to their inability to penetrate and recognize off-terrain objects, image matching techniques yield DEMs that 'are not very good, but acceptable for orthoimage production' (Gruen, 1996). Their main scope lies in small-scale applications. For use in runoff modelling the results in forested areas are not of acceptable quality.

Digitized contours as an elevation source are frequently used because of their relatively low costs. DEM generation is difficult from these data especially in areas of high contour curvature, such as valleys and ridges. These areas are most important for hydrology, since it is here that either channels or catchment borders are located. A particular problem is that general purpose interpolation techniques may yield flat areas or depressions in these zones (Rieger, 1992a). Some success has been achieved in the generation of consistent DEMs (Aumann *et al.*, 1991; Pilouk and Tempfli, 1992), but there are still inconsistencies because these algorithms use topological operations in areas of high contour curvature rather than fundamental geometric modelling. In flat areas, DEMs are generally weak owing to the arbitrary and coarse height resolution of contours. Yet, with proper constraints applied, the resulting DEMs at least provide consistent channel networks.

Time optimization and quality improvement techniques for photogrammetric DEM data collection are known as progressive, selective and composite sampling in (incomplete) grids (Makarovic, 1973, 1977; Charif, 1992) or at irregularly distributed points (Mann, 1988). One step may even be the extraction of skeleton information during the sampling process in order to improve data collection (Charif, 1992). This information can be used directly for hydrological applications. However, these techniques are not widely used, and if skeleton information is not digitized manually, but rather extracted from grid data, it again reflects the grid structure.

Representing the terrain surface through regular grids frequently leads to chains of depressions along narrow valleys, since only a few grid points lie on the valley bottom. Points close to the valley bottom frequently describe depressions. (Figure 2). The same is true for triangulated irregular network (TIN) DEMs, if the collected data points are regularly distributed, because usually the original data are used to define the TIN nodes. Most depressions are located in flat areas and in valley bottoms, especially in deep or narrow valleys (Rieger, 1992a).

Handling depressions

(a) *General.* In this discussion it is assumed that depressions are to be eliminated either explicitly or implicitly, so generating a 'depressionless DEM'. Flow can then be routed down to the DEM edges, with the exception of flagged (actual) depressions that are permitted to remain. Within valleys, depressions can normally be eliminated with little danger of gross errors, since runoff paths are well defined. For flat areas the situation is more difficult and gross misplacements of flow paths are possible.

Depressions can be eliminated at several stages. The optimum solution is that elevation data are collected in sympathy with the DEM generation process so that no artificial depressions are present. Alternatively, depressions can be eliminated during the DEM generation process (Hutchinson, 1989). However, in most cases depressions are still present after DEM generation and so they must be eliminated from it. Regardless of the stage at which the elimination process is included, it is necessary to understand and to take into account the reasons for the occurrence of depressions.

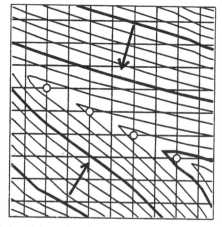

Figure 2. Chains of depressions in a narrow valley owing to grid structure

(b) *Methods to eliminate depressions*. The first step in eliminating a depression is to define the outlet point, usually the lowest point along the edge of the catchment border. Most of the available algorithms then raise the inner region of the depression. The following methods of achieving this have been described.

1. Raising the inner area to the elevation of the outlet point (O'Callaghan and Mark, 1984; Jenson and Domingue, 1988; McCormack *et al.*, 1993). The method has the great disadvantage that all form information within the depression is completely lost. Thus topological principles (rather than elevation information) are needed to define flow paths through the depression.

2. Rerouting the flow direction data set such that flow paths are defined (Chorowicz *et al.*, 1992). The principle is the same as in point 1 above, if routing is done merely topologically (Jenson and Domingue, 1988).

3. Raising the inner area and retaining the form of the depression (Rieger, 1992b). The inner area of the depression is raised between the elevations of the outlet point and its lowest outer neighbour, retaining the height relationships between the points. The flow path is then lowered from the depression to the outlet point. The advantage of the method is that the form is retained, so that there is no necessity to define flow directions arbitrarily within the depression.

4. Qian *et al.* (1990) and, similarly, Hadipriono *et al.* (1990) calculated drainage paths for a DEM with depressions, relating in a lot of unconnected pieces of drainage lines for which they derived a set of geometric properties. They then used expert system shells describing drainage network properties such as curvature, relative elevation of line pieces, or segment lengths, in order to find the most probable connections. However, the resulting channel networks did not correspond to reality as well as one might expect. So, either some parameters of the expert system shell (weights) need to be changed or some additional criteria need to be introduced.

5. Smoothing has sometimes been used to eliminate many small depressions (O'Callaghan and Mark, 1984; Tarboton *et al.*, 1990). This is, however, problematic since it flattens all curvature in a landscape and systematically shifts hillslopes.

None of the above methods takes into account the reasons for depressions. In fact, the methods of Qian *et al.* (1990) and Hadipriono *et al.* (1990) do not consider depressions explicitly. Chorowicz *et al.* (1992) come closest to what is suggested in the following discussion. As has been shown, depressions often exhibit correct terrain elevations compared with surrounding points that are too high, either because of systematically incorrect data or because of improper (arbitrary) point distributions. The following section provides a phenomenon-based approach that takes into account the main reasons for the occurrence of depressions.

(c) *A phenomenon-based approach*. Taking into account the main reasons for the occurrence of depressions, it appears that depression points are likely to reflect correct elevations, while the outlet points are too high. In these cases — particularly in valleys — phenomenon-based handling of depressions should prevent the elevation of the depressions from being lost. Rieger (1992b) suggested lowering the flow path from the depression, starting from the outlet point. The inner flow path leads to the deepest point of the depression, and the outer flow path to the first valley point that is deeper than the depression. The connecting path is lowered from the deepest point of the depression (inner end) to the outer end so that it leads monotonically downwards (Rieger, 1992b). The algorithm may also be applied to nested depressions by handling them in decreasing order (Figure 3).

FLOW ROUTING AND UPSLOPE CONTRIBUTING AREA

Definitions

Flow paths are usually defined from surface topography, but they may also be appropriate to ground-water flow (Anderson and Burt, 1978; Zaslavsky and Sinai, 1981; Bonell, 1993). Flow paths are used for flow

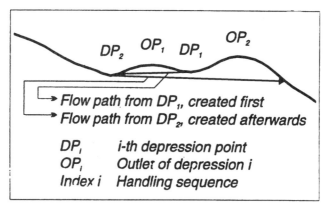

Figure 3. Lowering flow path in nested depressions. The figure shows a longitudinal valley profile

routing (Quinn *et al.*, 1991), delineation of catchments and estimation of upslope contributing areas (O'Callaghan and Mark, 1984; Jenson and Domingue, 1988; Rieger, 1993; and others), and, indirectly, for estimation of soil wetness indices (O'Loughlin, 1986).

There have been numerous attempts to estimate flow paths at the surface from DEMs, including grid models (O'Callaghan and Mark, 1984; Band, 1986; Jenson and Domingue, 1988; Skidmore, 1990; Quinn *et al.*, 1991; Rieger, 1992a,b; Tribe, 1992; McCormack *et al.*, 1993; Desmet and Govers, 1996; and others), TINs (Palacios-Vélez and Cuevas-Renaud, 1986) and contour DEMs (O'Loughlin, 1986; Moore, 1988; Moore and Grayson, 1989, 1991). Reviews are given by Moore *et al.* (1991, 1993), Tribe (1992), Beven and Moore (1993), and Desmet and Govers (1996). Here, only grid models are considered.

The total contributing area (TCA, also often denoted as A) is the plan area upslope of a surface element (normally, a contour line segment or grid cell) that drains to that element. The grid widths in the axis directions are denoted as Δx and Δy, respectively. In many cases a square grid is used, denoted by $\Delta = \Delta x$ and Δy.

Flow routing and slope direction

For contour models the calculation of flow paths is straightforward (O'Loughlin, 1986; Moore and Grayson, 1991). For TINs the ease of definition depends on the terrain position of the individual triangles. For grid models the situation is fairly complex, since one has either to leave the grid structure or to design algorithms that account for the arbitrary nature of the structure.

There are two classes of flow path algorithms: one-step algorithms estimate the TCA directly; two-step algorithms calculate an intermediate data set, the flow directions, and use it to integrate the TCA in the second step. The advantage of two-step algorithms is the generation of the flow direction data set which can be used for several applications (Jenson and Domingue, 1988). The disadvantage is that the generation of the flow direction data set requires a unique decision for each grid cell, as to which grid cell (or cells) flow occurs to. In contrast, one-step algorithms can be more sensitive to the actual slope direction.

Another common classification identifies single (sfd) and multiple flow direction (mfd) algorithms. Sfd algorithms route the accumulated TCA to one neighbouring grid cell, while mfd algorithms are able to split the TCA to several neighbours. The classical single flow algorithm (O'Callaghan and Mark, 1984), also known as 'steepest descent' and frequently named 'D8' (Costa-Cabral and Burges, 1994; Meijerink *et al.*, 1994; Blöschl, 1996) has widely been recognized as behaving poorly, since it yields slope lines parallel to grid lines or diagonals rather than parallel to the actual slope direction, and it is unable to model divergent flow appropriately (e.g. Fairfield and Leymarie, 1991; Rieger, 1992a,b; Tribe, 1992; Costa-Cabral and Burges, 1994).

There have been several attempts to improve this behaviour. Lea (1992) used an interpolated plane to define the gradient, but the preferential directions problem remains. Another approach switches randomly between deeper neighbours, resulting in a stochastically correct flow direction, called 'Rho4' or 'Rho8' for 4 or 8 neighbours, respectively (Fairfield and Leymarie, 1991). However, 'the re-introduction of some randomness into a model equation which previously has been freed from random influences ... appears as an illogical step backwards in the methodological research sequence' (Ahnert, 1994) and actually ignores available elevation information in the particular case.

Mfd algorithms normally split the TCA to all or several deeper neighbours, using a specific criterion. The principle is to disperse the flow in order to allow for directions other than the major grid directions. Mfd algorithms 'overspread' the TCA values, which is normally compensated for by the overspread TCA values of the neighbouring cells. Several approaches to splitting have been developed. Beasley *et al.* (1980) divide a cell by a line in the slope direction through a corner point. The two areas are routed to the neighbouring cardinal grid cells. The algorithm only allows the TCA to be split to two lower neighbours. Quinn *et al.* (1991) and Wolock and McCabe (1995) routed the TCA weighted by the elevation differences to the lower of its eight neighbours, referred to as 'DH8'. The same approach has been used for only four neighbours by Rieger (1992a,b; 1993), referred to as 'DH4'. Quinn *et al.* (1995) have modified DH8 by an exponent to the slope. Other researchers have derived local curvature and combined the resulting pieces of the drainage network in several ways (Band, 1986; Chorowicz *et al.*, 1992), and expert systems that use information from several geometric properties have been employed by Qian *et al.* (1990) and Hadipriono *et al.* (1990). Costa-Cabral and Burges (1994) were the first to model slope lines between grid cells ('flowing tubes'). Flow is routed from each cell upwards [in order to compute TCA and specific upslope contributing area (SCA)] and downwards [to compute SDA (specific dispersal area) and SCA] along all possible paths. Within each cell, the portion of the respective flow path is modelled as a band parallel to the slope direction of that cell.

A key question is the number of neighbours involved in the routing process. For sfd algorithms normally the eight surrounding neighbours are used, allowing for the best resolution of direction. For mfd algorithms the question is more complex. Geometrically, only the flux to the four cardinal neighbours is defined. Nevertheless, it is common to route to eight neighbours even with a square grid, since this allows for a more sensible flow distribution.

Several studies have compared the results of sfd and mfd algorithms with one another (Quinn *et al.*, 1991; Wolock and McCabe, 1995; Desmet and Govers, 1996). The mfd algorithms tend to yield smoother distributions of TCA values than the sfd algorithms. Costa-Cabral and Burges (1994) used synthetic surfaces, which allows comparison of the results with correct data. The same approach is adopted here.

One aim of flow algorithms is to estimate slope lines. As mentioned, sfd algorithms behave poorly except for the Rho8 algorithm (Fairfield and Leymarie, 1991). Mfd algorithms spread the area unrealistically and do not readily provide a single downslope path or connected flow direction data set, as can be created with sfd algorithms. They can, however, be used to estimate slope paths for single cells (Rieger, 1992b). Initializing only the source cell with a non-zero value, the flow algorithm yields a field of dispersion values that can easily be traced downwards, always stepping to the lower neighbour with the maximum dispersion value. The result accurately represents the flow path for those algorithms that spread the area proportional to elevation differences as shown for a tilted plane with the DH4 (Figure 4) and DH8 algorithm (Figure 5). The area-weighted algorithm of Beasley *et al.* (1980) and a modified version of DH8, which spreads proportionally to a power (other than one) of slope (Quinn *et al.*, 1995), do not show the same property.

Accurate modelling of flow paths is only possible either by modelling subgrid flow channels (Costa-Cabral and Burges, 1994) or by mfd algorithms that distribute the upslope area such that the slope lines are represented, which is the case for both the DH4 and the DH8 algorithm. All these methods can be used to delineate single flow paths to specific seed points. The generation of a flow direction data set is also possible, yet very time-consuming. Flow paths need to be traced from all points downwards in order to define appropriate flow directions. For each point a flow accumulation data set must be initialized with zeroes, and a seed value has to be stored at the point's position. The seed value is routed downwards (in ordered sequence; Rieger, 1992a,b). Finally, the flow path is traced until it reaches an already recorded flow direction value or the edge of the DEM, and the flow directions are stored in the flow direction data set. The whole process is

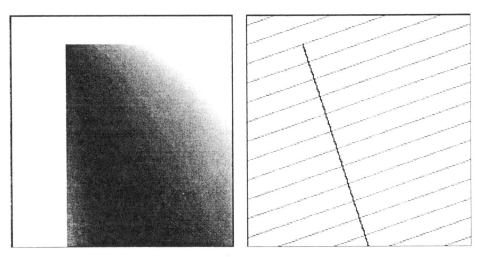

Figure 4. DH4 algorithm: dispersion of a water unit in a seed point on a tilted plane. (Left) Log of dispersion values (DV), black: DV = 1; white: DV = 0. (Right) Line of downslope maximum dispersion values crosses contours at a right angle

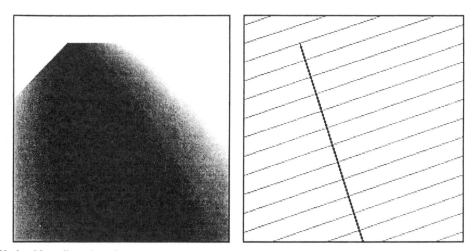

Figure 5. DH8 algorithm: dispersion of a water unit in a seed point on a tilted plane. (Left) Log of dispersion values (DV), black: DV = 1; white: DV = 0. (Right) Line of downslope maximum dispersion values crosses contours at a right angle

done in sequence of decreasing elevation for all points that do not yet have flow directions assigned. The resulting flow direction data set better reflects the DEM geometry than all known sfd algorithms.

The total upslope contributing area (TCA)

The TCA is the area that drains to a given surface element, the 'reference element'. Although TCA values have been estimated in several different ways, little attention has been given to the significance of the associated surface element. For DEMs one would normally assume an element of the DEM structure (grid cell in grid models, triangle in TINs). Flow algorithms need to be designed such that the resulting TCA values reflect the reference element size.

For the D8 (and other) sfd algorithm the TCA values correspond to a reference element length of grid width, Δ, when flow occurs in a cardinal (axis) direction, and of $\Delta/2$ in a diagonal direction (e.g. Figure 6). Since there are no other flow directions possible, no other reference element widths occur. In zones of convergence (valleys) the TCA is well represented and fairly direction independent, since the sfd algorithms

a	10	10	10
	11	11	11
	12	12	12
	13	13	13

b	10	10	11	11	12	12
	10	11	11	12	12	13
	11	11	12	12	13	13
	11	12	12	13	13	14

Figure 6. D8 algorithm: TCA values on a plane with a slope direction (a) towards the base of the diagram (negative y-direction, (b) towards the bottom right of the diagram (positive x- and negative y-direction). $\Delta x = \Delta y = 1$

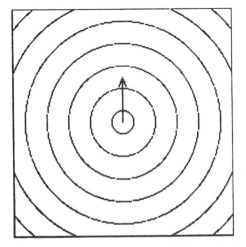

Figure 7. Contours of a vertical cone with its peak pointing upwards (out of the page). The arrow points downhill

Figure 8. DH4 algorithm: lines of equal TCA values are plotted on a cone with its peak pointing upwards (out of the page)

simply count the upslope cells; so the correctness of the results mainly depends on the correctness of the catchment borders.

For mfd algorithms the situation is more complex. The number of possible directions is not discrete, so there is a continuously changing relation to the aspect angle. Figure 7 shows the contour lines of a cone with a vertical axis and upwards peak. Figure 8 shows lines of equal TCA values for the DH4 algorithm on that cone. These lines exactly correspond to TCA values that are defined for the width of a grid cell seen in aspect direction α. With grid widths Δx and Δy, the width b_α of a grid cell viewed at a certain aspect angle α can be expressed as (Figure 9),

$$b_\alpha = \Delta x \cdot |\cos\alpha| + \Delta y \cdot |\sin\alpha| \tag{1}$$

or, with equal grid sizes in both axis directions, $\Delta x = \Delta y = \Delta$, and $b_0 = \Delta$ as reference element size,

$$b_\alpha = \Delta \cdot (|\cos\alpha| + |\sin\alpha|) = b_0 \cdot (|\cos\alpha| + |\sin\alpha|) \tag{2}$$

The scaling factor in Equation (2) varies between $1 \cdot 0$ in the axis direction ($\alpha = 0, \pi/2, \pi, \ldots$) and the square root of 2 ($1 \cdot 414$) in the diagonal direction, ($\alpha = \pi/4, 3\pi/4, \ldots$). The theoretical TCA to a contour piece b on a cone at a distance R from the centre point is calculated according to (Figure 10),

$$A = b \cdot R/2 \tag{3}$$

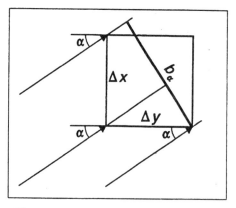

Figure 9. Width of grid cell seen in direction α

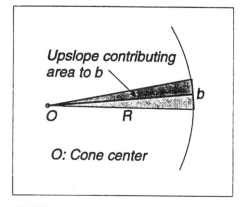

Figure 10. TCA to contour line segment b at a distance R from the centre point of a cone with its peak pointing upwards (out of the page)

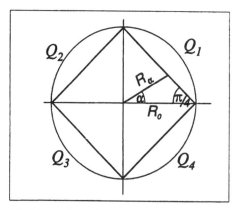

Figure 11. Distance from the centre point to a straight diagonal line in comparison with a circle

resulting in circular lines of equal TCA values that increase linearly with distance from the centre point. Changing the reference element size, b, to b_α as above, yields direction-dependent TCA values at a constant distance R

$$A_\alpha = b_\alpha \cdot R/2 \tag{4}$$

The lines of equal TCA values in that case are concentric squares rather than circles, rotated by $\pi/4$, since the distance R_α of any point along the edge of such a square depends on the direction (Figure 11),

$$R_\alpha = R_0 \cdot \frac{\sin\pi/4}{\sin(\pi/4 + \alpha)} = R_0 \cdot \frac{1}{\cos\alpha + \sin\alpha} \tag{5}$$

for example, in quadrant Q_1. Generalization to the other three quadrants introduces the absolutes of sine and cosine, resulting in the same factors as between b and b_α [Equation (2)]. The DH4 algorithm produces the same results, and so normalization takes place by division of the calculated TCA values by the factor of Equation (2),

$$A_{\mathrm{norm}} = A/(|\cos\alpha| + |\sin\alpha|) \tag{6}$$

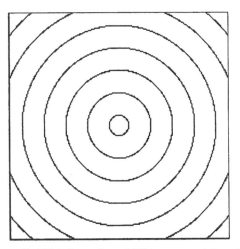

Figure 12. DH4: lines of equal TCAs, corrected for grid
direction, on an upward pointing cone

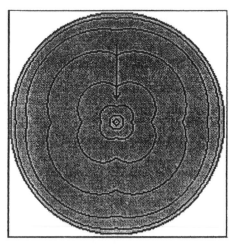 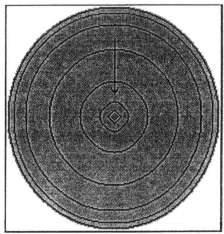

Figure 13. DH4: cone with peak pointing downwards (into the page). (Left) lines of equal log of TCA. (Right) standardized for grid
direction. Only the grey area is a part of the DEM

Figure 12 shows the normalized TCA values of the cone of Figure 7, which are exact circles. Figure 13 shows
results for DH4 on a cone with its top pointing down. The corrected values exhibit systematic deviations
from the theoretical TCA values close to the centre point.

Figure 14 shows TCA-values for the DH8 algorithm on a cone with its peak pointing up. Figure 15
illustrates the same algorithm applied to a cone with its peak pointing down. The direction dependence is
less marked than with the DH4, especially in the convergent case (Figure 15 *vs.* Figure 13). For the DH8
algorithm the correction factor is more complex; it corresponds to the width of an octagonal pseudo-grid cell
seen in direction α, and is similar to Equation (5), with $3\pi/8$ instead of $\pi/4$

$$b_{\alpha,8} = b_0 \cdot \frac{\sin\left(\dfrac{3\pi}{8} + \alpha^*\right)}{\sin\dfrac{3\pi}{8}} = b_0 \cdot \left[\left(\sqrt{2} - 1\right) \cdot \sin\alpha^* + \cos\alpha^*\right] \qquad (7)$$

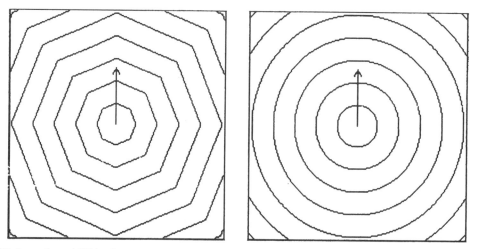

Figure 14. DH8: lines of equal TCA values on an upward pointing cone. (Left) uncorrected. (Right) standardized for grid direction

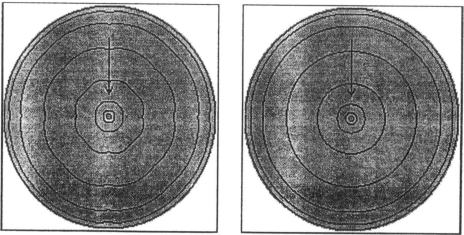

Figure 15. Cone with peak pointing downwards. (Left) lines of equal log of TCA. (Right) standardized for grid direction. Only the grey area is part of the DEM

where α^* denotes angle α reduced to the interval $[0 \ldots \pi/4]$

$$\alpha^* = \alpha - \left[\frac{4\alpha}{\pi}\right] \cdot \frac{\pi}{4} \tag{8}$$

with $[x]$ meaning the highest integer value that is lower than or equal to x. Quinn *et al.* (1995) use a different approach, distributing a TCA value to cell i by

$$A_i = A_0 \cdot \frac{(\tan\beta_i)^p}{\sum\limits_j (\tan\beta_j)^p} \tag{9}$$

with an exponent $p = 1 \cdot 1$. Figure 16 shows the corresponding lines of equal TCA on cones with their peaks pointing up and down, respectively. The direction dependency is reduced but is still present. The slope direction on a tilted plane, when calculated as line of maximum TCA values downwards from a seed point, is slightly wrong.

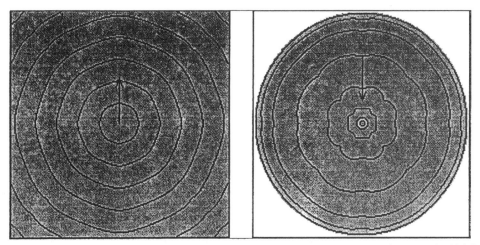

Figure 16. DH8 with exponent $p = 1.1$. Lines of equal log of TCA on a cone (left) with its top pointing upwards; (right) with its top pointing downwards. Only the grey area is part of the DEM

For the simple case of cones or tilted planes, a scaling factor can be applied in order to obtain direction independent results. However, with any change in direction along any flow path the length of the reference contour element changes within the extremes of b_α, resulting in erroneous TCA values. The error in these estimates attributable to slope direction depends on the flow algorithm used. Table I summarizes the maximum errors of several flow algorithms, compared with a contour length of grid width Δ. When TCA values are corrected by the factor of Equation (7), the remaining deviations resulting from grid direction of the DH8 algorithm will normally lie within $\pm 4\%$ of the actual TCA values.

As an alternative to the grid-dependent calculation of the TCAs, a flow direction data set can be used. If it has been calculated as described above, a fairly direction-independent estimate for the TCAs can be obtained by simply counting the number of grid cells upslope of the reference element (e.g. Jenson and Domingue, 1988). In this case, the flow algorithms are used to create the flow direction data set rather than to estimate the TCAs. To obtain an even more accurate measure, the 'catchment affiliation values' can be used (Rieger, 1992b). These values result from applying the flow routing algorithm in an upslope direction and yield the percentage of area in any cell that is routed to a given catchment area. The area of cells along the border of two catchments can be split according to these values, while the inner areas of the catchments are estimated by the counting process.

One problem that has not been dealt with in the above considerations is the assumption of linearity between the TCA and the contour length

$$TCA(k \cdot b) = k \cdot TCA(b) \qquad (10)$$

This relation is strictly fulfilled only on special surfaces, such as tilted planes or cones. If linearity is not given, the application of a scaling factor on the TCA values will not yield correct results. However, the influence will

Table I. Extreme deviations of TCA owing to slope direction for several flow algorithms on a plane, compared with a reference element of grid width. The dots ('...') denote a continuous change in intermediate directions

Code	Description	Type	Number of neighbours	Correction factor b_0/b_α		Deviation from actual TCA (%)
D8	Steepest descent	sfd	8	1·000	1·414	0, −29
DH4	Δh weighted	mfd	4	1·000	... 0·707	0 ... + 41
DH8	Δh weighted	mfd	8	1·000	... 0·924	0 ... + 8

normally be small. On *hillslopes*, significant changes in flow direction are rare; *convex areas* are normally of minor interest, since the TCAs are small, resulting in low erosion and soil wetness as well as in small absolute errors of the TCAs; within *concave areas* (valleys) the linearity criterion is likely not to be fulfilled and the influence of the local flow direction (i.e. valley direction) is small compared with the much larger areas of surrounding hillslopes, which may exhibit very different flow directions. In these areas the DH8 algorithm will yield the best results.

CONCLUSION

Given the main causes of depressions in grid DEMs, raising the inner areas is frequently inappropriate as a method of depression removal, since depression points, particularly in valleys, tend to reflect the channel position better than the surrounding points. In these cases, cutting flow channels is suggested as an alternative.

The effects of the grid direction on the estimation of flow paths have also been evaluated. The DH8 algorithm has been shown to behave best in terms of rotation invariance and correctness of flow direction. The application of a direction-dependent scaling factor further improves the TCA values. Further research is needed on how changes in upslope flow direction affect the TCA, especially in valleys.

REFERENCES

Ackermann, F. 1992. 'Experience with automatic DEM generation', *Proc. 17th ISPRS Congress, Washington, D.C., Comm. IV, IAPRS*, **28**(B4), 986–989.

Ahnert, F. 1994. 'Randomness in geomorphological process response models', in Kirkby, M. J. (ed.), *Process Models and Theoretical Geomorphology*. John Wiley & Sons, Chichester. pp. 3–21.

Anderson, M. G. and Burt, T. P. 1978. 'The role of topography in controlling throughflow generation', *Earth Surf. Process Landf.*, **3**, 331–344.

Aumann, G., Ebner, H., and Tang, L. 1991. 'Automatic derivation of skeleton lines from digitized contours', *ISPRS J. Photogramm. Remote Sens.*, **46**, 259–268.

Baltsavias, E. P., Li, H., Stefanidis, A., Sinning, M., and Mason, S. 1996. 'Comparison of two digital photogrammetric systems with emphasis on DTM generation: case study glacier measurement', *Proc. 18th ISPRS Congress, Vienna, Comm. IV/4, IAPRS*, **31**(B4), 104–109.

Band, L. E. 1986. 'Topographic partitioning of watersheds with digital elevation models', *Wat. Resour. Res.*, **22**, 15–24.

Beasley, D. G., Huggins, L. F., and Monke, E. J. 1980. 'ANSWERS — a model for watershed planning', *Trans. ASAE.* **23**, 938–946.

Beven, K. J. and Moore, I. D. 1993. *Terrain Analysis and Distributed Modelling in Hydrology*. John Wiley & Sons, Chichester.

Blöschl, G. 1996. *Scale and Scaling in Hydrology*. John Wiley, Chichester, in press.

Bonell, M. 1993. 'Progress in the understanding of runoff generation dynamics in forests', *J. Hydrol.*, **150**, 217–275.

Charif, M. 1992. 'Optimum sampling for digital terrain modelling'. *Proc. 17th ISPRS Congress, Washington, D.C., Comm. III, IAPRS*, **28**(B3), 77–86.

Chorowicz, J., Ichoku, C., Riazanoff, S., Kim, Y.-J., and Cervelle, B. 1992. 'A combined algorithm for automated drainage network extraction', *Wat. Resour. Res.*, **28**, 1293–1302.

Costa-Cabral, M. C. and Burges, S. J. 1994. 'Digital elevation model networks (DEMON): a model of flow over hillslopes for computation of contributing and dispersal areas', *Wat. Resour. Res.*, **30**, 1681–1692.

Desmet, P. J. J. and Govers, G. 1996. 'Comparison of routing algorithms for digital elevation models and their implications for predicting ephemeral gullies', *Int. J. GIS*, **10**, 311–331.

Fairfield, J. and Leymarie, P. 1991. 'Drainage networks from grid digital elevation models', *Wat. Resour. Res.*, **27**, 709–717.

Freeze, R. A. and Harlan, R. L. 1969. 'Blueprint for a physically-based, digitally-simulated hydrologic response model', *J. Hydrol.*, **9**, 237–258.

Gruen, A. 1996. 'Digital photogrammetric stations revisited', *Proc. 18th ISPRS Congress, Vienna, Comm. II/III, IAPRS*, **31**(B2), 127–132.

Hadipriono, F. C., Lyon, J. G., Li, T. W. H., and Argialas, D. P. 1990. 'The development of a knowledge-based expert system for analysis of drainage patterns', *Photogramm. Eng. Remote Sens.*, **56**, 905–909.

Hahn, M. and Förstner, W. 1988. 'The applicability of a feature based matching and a least squares matching algorithm for DTM acquisition, *Proc. 16. ISPRS Congress, Kyoto, Comm. III, IAPRS*, **27**(B9), 137–150.

Helava, U. V. 1988. 'Object space least squares correlation', *Proc. 16. ISPRS Congress, Kyoto, Comm. II, IAPRS*, **27**(B2), 297–302.

Hutchinson, M. F. 1989. 'A new procedure for gridding elevation and stream line data with automatic removal of spurious pits', *J. Hydrol.*, **106**, 211–232.

Jenson, S. K. and Domingue, J. O. 1988. 'Extracting topographic structure from digital elevation data for GIS analysis', *Photogramm. Eng Remote Sens.*, **54**, 1593–1600.

Kaiser, B., Hausladen, J., Tsay, J. R., and Wrobel, B. P. 1992. 'Application of FAST vision for digital terrain model generation', *Proc. 17th ISPRS Congress, Washington, D.C., Comm. IV, IAPRS*, **28**(B4), 809–816.

Kilian, J., Haala, N., and Englich, M. 1996. 'Capture and evaluation of airborne laser scanner data', *Proc. 18th ISPRS Congress, Vienna, Comm. III/1, IAPRS*, **31**(B3), 383–388.

Kraus, K. 1984. *Photogrammetrie*, Band II. Dümler Verlag, Bonn; In English *Photogrammetry*, Vol. II, 1997.

Kraus, K., Reiter, T., Hynst, E., and Belada, P. 1997. 'Data in wooded regions', *Proc. of the Third Joint European Conference and Exhibition on Geographical Information, April 16–18, Vienna, Austria*, Vol. 2. pp. 966–975.

Krzystek, P. 1991. 'Fully automatic measurement of digital elevation models with MATCH-T', *Proc. 43rd Photogrammetric Week, Stuttgart. Inst. f. Photogrammetric, Universität Stuttgart*. pp. 203–214.

Kubik, K. 1988. 'Digital elevation models. Review and outlook', *Proc. 16. ISPRS Congress, Kyoto, Comm. III, IAPRS*, **27**(B3), 415–426.

Lea, N. L. 1992. 'An aspect driven kinematic routing algorithm', in Parsons, A. J. and Abrahams, A. D. (eds), *Overland Flow: Hydraulics and Erosion Mechanics*. Chapman and Hall, New York.

Lohr, U. and Eibert, M. 1995. 'The TopoSys laser scanner-system', in Fritsch, D. and Hobbie, D. (eds), *Proceedings of the Photogrammetric Week 1995, Stuttgart*. Wichmann, Heidelberg. pp. 263–268.

Makarovic, B. 1973. 'Progressive sampling for DTM's', *ITC-J*, **1973-3**, 397–416.

Makarovic, B. 1977. 'Composite sampling for digital terrain models', *ITC-J*, **1977-3**, 406–431.

Mann, H. 1988. 'Progressive sampling', *Proc. 16. ISPRS Congress, Kyoto, Comm. II, IAPRS*, **27**(B10), 319–325.

Mark, D. M. 1979. 'Phenomenon-based data-structuring and digital terrain modelling', *Geoprocessing*, **1**, 27–36.

Mark, D. M., Dozier, J., and Frew, J. 1984. 'Automated basin delineation from digital elevation data', *Geoprocessing*, **2**, 299–311.

McCormack, J. F., Gahegan, M. N., Roberts, S. A., Hogg, J., and Hoyle, B. S. 1993. 'Feature-based derivation of drainage networks', *Int. J. GIS*, **7**, 263–279.

Meijerink, A. M. J., de Brouwer, H. A. M., Mannaerts, C. M., and Valenzuela, C. R. 1994. *Introduction to the Use of Geographic Information Systems for Practical Hydrology*, ITC Publication No. 23, ITC, Enschede. 243 pp.

Miller, S. B. and DeVenecia, K. 1992. 'Automatic elevation extraction and the digital photogrammetric workstation', *Proc. ASPRS-ACSM Annual Convention, Albuquerque, ASPRS Technical Papers*, Vol. 1. pp. 572–580.

Montgomery, D. R. 1996. 'Reply', *Wat. Resour. Res.*, **32**, 1463–1465.

Moore, I. D. 1988. 'A contour-based terrain analysis program for the environmental sciences (TAPES)', *Trans. Am. Geophys. Union*, **69**, 345.

Moore, I. D. and Grayson, R. B. 1989. 'Hydrologic and digital terrain modelling using vector elevation data', *Trans. Am. Geophys. Union*, **70**, 1091.

Moore, I. D. and Grayson, R. B. 1991. 'Terrain-based catchment partitioning and runoff prediction using vector elevation data', *Wat. Resour. Res.*, **27**, 1177–1191.

Moore, I. D., Grayson, R. B., and Ladson, A. R. 1991. 'Digital terrain modelling: a review of hydrological, geomorphological, and biological applications', *Hydrol. Process.*, **5**, 3–30.

Moore, I. D., Lewis, A., and Gallant, J. C. 1993. 'Terrain attributes: estimation methods and scale effects', in Jakeman, A. J., Beck, M. B., and McAleer, M. (eds), *Modelling Change in Environmental Systems*. Wiley, Chichester. pp. 189–214.

Næsset, E. 1997. 'Determination of mean tree height of forest stands using airborne laser scanner data', *ISPRS J. Photogramm. Remote Sens.*, **52**, 49–56.

O'Callaghan, J. F. and Mark, D. M. 1984. 'The extraction of drainage networks from digital elevation data', *Comp. Vis. Graph. Image Proc.*, **28**, 323–344.

O'Loughlin, E. M. 1986. 'Prediction of surface saturation zones in natural catchments by topographic analysis', *Wat. Resour. Res.*, **22**, 794–804.

Palacios-Vélez, O. L. and Cuevas-Renaud, B. 1986. 'Automated river-course, ridge, and basin delineation from digital elevation data', *J. Hydrol.*, **86**, 299–314.

Pilouk, M. and Tempfli, K. 1992. 'A digital image processing approach to creating DTMs from digitized contours', *Proc. 17th ISPRS Congress, Washington, D.C., Comm. IV, IAPRS*, **28**(B4), 956–961.

Qian, J., Ehrich, R. W., and Campbell, J. B. 1990. 'DNESYS — an expert system for automatic extraction of drainage networks from digital elevation data', *IEEE Trans. Geosci. Remote Sens.*, **28**, 29–45.

Quinn, P. F., Beven, K. J., Chevallier, P., and Planchon, O. 1991. 'The prediction of hillslope flow paths for distributed hydrological modelling using digital terrain models', *Hydrolog. Process.*, **5**, 59–79.

Quinn, P. F., Beven, K. J., and Lamb, R. 1995. 'The $\ln(a/\tan\beta)$ index: how to calculate it and how to use it within the TOPMODEL framework', *Hydrol. Process.*, **9**, 161–182.

Rieger, W. 1992a. 'Hydrologische Anwendungen des digitalen Geländemodelles', *Dissertation*, Technical University of Vienna.

Rieger, W. 1992b. 'Automated river line and catchment area extraction from DEM data', *Proc. 17th ISPRS Congress, Washington, D.C., IAPRS*, **28**(B4), 642–649.

Rieger, W. 1993. 'Hydrological terrain features derived from a pyramidal raster structure', *Proc. HydroGIS '93 (Application of GISs in Hydrology and Water Res.), Vienna, IAHS Publ.*, **211**, 201–210.

Skidmore, A. K. 1990. 'Terrain position as mapped from a digital elevation model', *Int. J. GIS*, **4**, 33–49.

Tarboton, D. G., Bras, R. L., and Rodríguez-Iturbe, I. 1991. 'On the extraction of channel networks from digital elevation data', *Hydrol. Process.*, **5**, 81–100.

Tribe, A. 1992. 'Automated recognition of valley lines and drainage networks from grid digital elevation models: a review and a new method', *J. Hydrol.*, **139**, 263–293.

Walker, A. S. and Petrie, G. 1996. 'Digital photogrammetric workstations 1992–96', *Proc. 18th ISPRS Congress, Vienna, Comm. II/III, IAPRS,* **31**(B2), 384–395.

Wild, D. and Krzystek, P. 1996. 'Automatic breakline detection using an edge preserving filter', *Proc. 18th ISPRS Congress, Vienna, Comm. III/3, IAPRS,* **31**(B3), 946–952.

Wilson, J. P. 1996. 'GIS-based land surface/subsurface modeling: new potential for new models?', *Third International Conference/ Workshop on Integrating GIS and Environmental Modeling, Santa Fe, New Mexico, Jan. 21–25, 1996. Proc. on CD-Rom, NCGIA, Univ. of California, Santa Barbara.*

Wolock, D. M. and McCabe, Jr, G. J. 1995. 'Comparison of single and multiple flow direction algorithms for computing topographic parameters in TOPMODEL', *Wat. Resour. Res.,* **31**, 1315–1324.

Zaslavsky, D. and Sinai, G. 1981. 'Surface hydrology: I. Explanation of phenomena, II. Distribution of raindrops, III. Cause of lateral flow, IV. Flow in sloping layered soil, V. In-surface transient flow', *Proc. Am. Soc. Civil Eng., J. Hydraul.,* **107**(HY1), 1–93.

Zhang, W. and Montgomery, D. R. 1994. 'Digital elevation model grid size, landscape representation, and hydrologic simulations', *Wat. Resour. Res.,* **30**, 1019–1028.

5

LARGE-SCALE DISTRIBUTION MODELLING AND THE UTILITY OF DETAILED GROUND DATA

FRED G. R. WATSON,[1]* RODGER B. GRAYSON,[1] ROBERT A. VERTESSY [2] AND THOMAS A. McMAHON[1]

[1] *Cooperative Research Centre for Catchment Hydrology, Department of Civil and Environmental Engineering, University of Melbourne, Parkville, Victoria 3052, Australia*
[2] *Cooperative Research Centre for Catchment Hydrology, CSIRO Land and Water, GPO Box 1666, Canberra, ACT 2601, Australia*

ABSTRACT

A large-scale distribution function model was used to investigate the effect of differing parameter mapping schemes on the quality of hydrological predictions.

Precipitation was mapped over a large forested catchment area (163 km^2) using both one-dimensional linear and three-dimensional non-linear interpolation schemes. Lumped stream flow predictions were found to be particularly sensitive to the different precipitation maps, with the three-dimensional map predicting 12% higher mean annual precipitation, resulting in 36% higher modelled stream flow over a three-year period. However, spatial predictions of stream flow appeared worse when derived from the three-dimensional map, which is considered the better of the two precipitation maps. This implies uncertainty in either the model's response to precipitation or the precipitation mapping process (the 12% precipitation difference was strongly determined by a single, short term gauge). Leaf area index (LAI) was mapped using both remote sensing and species based methods. The two LAI maps had similar lumped mean values but exhibited significant spatial differences. The resulting lumped predictions of stream flow did not vary. This suggests a linear response of water balance to LAI in the non-water-limited conditions of the study area, and de-emphasizes the importance of quantifying *relative* spatial variations in LAI. Topographic maps were created for a small experimental subcatchment (15 ha) using both air photographic interpretation and ground survey. The two maps differed markedly and lead to significantly different spatial predictions of runoff generation, but nearly identical predicted hydrographs. Thus, *at scales of small basins*, accurate topographic mapping is suggested to be of little importance in distribution function modelling because models are unable to make use of complex spatial data.

Predictions of water yield can be very sensitive (in the case of precipitation) or insensitive (in the case of small-scale topography) to changes in spatial parameterization. In either case, increased complexity in spatial parameterization does not necessarily result in better, or more certain prediction of hydrological response.

KEY WORDS   forest hydrology; distribution function modelling; parameterization; uncertainty; precipitation; leaf area index (LAI); topography; digital elevation models (DEM); Macaque

INTRODUCTION

The realism of large-scale spatial models (LSSMs)

Our physical understanding of catchment hydrological processes is founded on observations at scales of 0·1 to 100 metres. Large-scale spatial models (LSSMs) deal with catchments much larger than this scale

---

*Correspondence to: Fred G. R. Watson, Cooperative Research Centre for Catchment Hydrology, Department of Civil and Environmental Engineering, University of Melbourne, Parkville, Victoria 3052, Australia.

and must broach the gap in scales by dividing catchments into many smaller spatial units, each of which represents an area within which physical processes are simulated. The characterization of a given spatial unit is made through a number of input parameters which must be mapped across the catchment so that each spatial unit may be modelled correctly. We view this as a two-goal system. The first goal is the representation of detailed processes within an elementary spatial unit. The second goal is the division of the catchment into appropriate elementary spatial units, and the mapping of the parameters that distinguish and control the individual behaviour of those units throughout the catchment. The achievement of a *realistic* simulation depends on both goals being attained. It is questionable whether any LSSM modelling project to date has simultaneously attained both goals. This paper investigates the second goal. Specifically, the effects of differing parameter mapping strategies on model behaviour are tested. The degree to which the ensuing results influence the uncertainty in model realism is discussed.

Outline of research

The wider aims of our work are to investigate the effects of land cover change and water yield in the water supply catchments of the city of Melbourne in south-east Australia (Figure 1). Ultimately, we aim to predict long-term (*c.* 100 years) changes in water yield associated with possible future changes in land cover; and to understand the physical processes governing water balance and its change (see Watson *et al.*, 1997).

The specific aims of this paper are: to investigate how differences in parameter mapping strategies affect spatial and lumped predictions of water yield made by spatial models, in this case, Macaque; and to observe whether added spatial complexity in parameterization improves model predictions and/or decreases their uncertainty. The conclusions we draw are applicable to most LSSMs.

The water balance of the catchments in question is hypothesized to depend on (amongst other things) precipitation, forest leaf area index (LAI) and topography. Precipitation and LAI vary markedly in both space and time throughout the catchments and topography varies markedly in space. A spatial model provides a framework within which we may organize our enquiry and understanding of the interactions between these factors as they vary in space and time. A spatial model also allows the simulation of future land cover conditions.

Thus, the spatial mapping of three parameters is investigated

- precipitation,
- leaf area index (LAI), and
- topography.

Initially, maps of these parameters are produced using commonly available methods. The model is parameterized and calibrated using these maps to the point where it operates in an apparently realistic manner. Then, each parameter in turn is mapped by some alternative means based on a different type of spatial information, and the model is run using this map. Differences in model operation are assessed by comparing spatial and temporal stream flow predictions.

In the remainder of this paper, we briefly describe the model, and then introduce the study area and the initial application of Macaque to the study area. In the later sections, different parameter mapping strategies are investigated in turn with respect to their effect on simulated water yield before a discussion of all the results is presented.

MODEL DESCRIPTION

Macaque is a large-scale, long-term, physically based water balance model which predicts the water yield of forested catchments subject to land cover change. It operates at a daily time-step and is designed to simulate water balance during periods of over 100 years for catchments over 100 km^2 disaggregated into over 1000 elementary spatial units (ESUs). In this special issue on GIS applications in hydrology, space does not

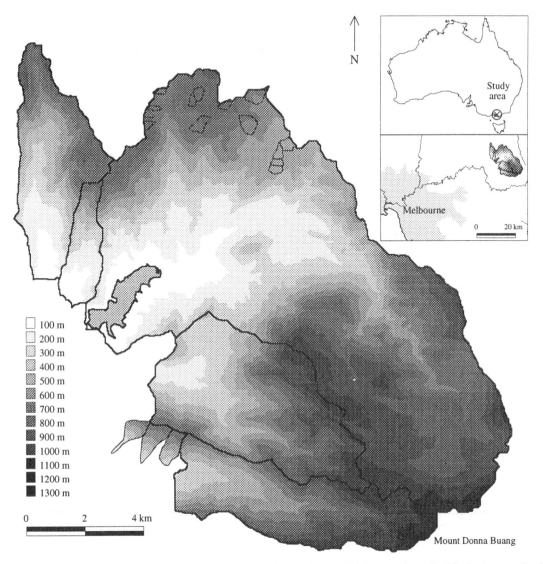

Figure 1. Digital elevation model and location of the study area 55 km north-east of Melbourne, Australia. The five large outlined areas are water supply catchments and the smaller outlined areas are experimental catchments

permit a full description of this new model. Our focus is on the use of spatial data in spatial models rather than on the model itself. A summary description of Macaque is, however, presented, highlighting the key differences from parent models. The reader is referred to Watson (1997) for further details.

Macaque is a distribution function model (DFM), operating in a similar way to RHESSys (Band et al., 1993) and TOPMODEL (Beven et al., 1994). Catchments are first divided into a number of hillslopes. Then, a topographic wetness index is calculated for all parts of each hillslope and areas of similar wetness index are grouped together as the ESUs of the model. Lateral flow between ESUs is implemented implicitly using a generalized distribution function whereby, at each time-step, total hillslope saturated zone water is re-distributed amongst the ESUs in proportion to their wetness index and the mean wetness index of the hillslope.

A key difference between Macaque and other DFMs is that Macaque is what we term a *limited* DFM. This involves the limiting of lateral re-distribution of water according to a *lateral redistribution factor*. This can be thought of as a quantification of subsurface lateral hydraulic conductivity. The lateral redistribution factor allows local variations in water table level to develop according to local recharge and evapotranspiration conditions. An example of the utility of this feature is as follows. The study area described below exhibits deep soils and deep water tables which only approach the surface near streams. Throughout most of the area, the water table depth is relatively static, but near the streams, proximity to surface processes causes the water table to rise and fall with individual storms. Surface saturated areas respond to the same dynamic. Conventional DFMs (such as TOPMODEL) force all water table variations to be homogeneous, to occur in unison throughout the hillslope. The limited DFM approach slows down the lateral redistribution of water and allows local water table variations to develop, permitting realistic simulation of near-stream water table and surface saturated area dynamics. The approach, whilst conceptual in nature, has been validated (dynamically) using daily water table data from a transect of piezometers. A related difference between Macaque and other DFMs is that baseflow is calculated independently *for each ESU* according to its area and the degree to which it is saturated. This forces an explicit representation of baseflow generation from dynamic saturated areas.

Within each ESU, the model implements a three-layer (canopy, understorey and soil) Penman–Monteith representation of evapotranspiration, with detailed representation of such factors as energy balance, leaf conductance, interception storage, radiation propagation, humidity gradients and soil water extraction. Much of the vertical structure is based on that embodied within Topog (Vertessy *et al.*, 1993, 1995b; Hatton *et al.*, 1995), COUPLES (Silberstein and Sivapalan, 1995) and RHESSys (Band *et al.*, 1993).

From the software engineering perspective, the spatial data structure and algorithms of Macaque are very highly refined. Features include: recursive, generic code for simulating different levels of spatial disaggregation (catchments, hillslopes, ESUs, etc.); hierarchical parameter inheritance between these levels; the ability to specify the value of any parameter at any spatial level at run time; and automatic spatio-temporal accounting and output of any model variable (including all inputs, outputs, stores, parameters and intermediate working variables). The code is completely original.

The model uses a large number of parameters, most of which are held constant in space and time. Generally, the climatic and topographic parameters are easily quantified whilst vegetation and soil parameters vary from being relatively insensitive and easily measured (e.g. albedo, porosity) to being quite sensitive and difficult to fix (e.g. hydraulic conductivity of the soil, aerodynamic resistance of the canopy to water vapour transfer). A number of key parameters are varied in space and time (e.g. LAI) and are the focus of this paper. Time-series inputs are limited to daily precipitation and maximum and minimum temperature at a base station.

THE STUDY AREA

The study area comprises five water supply catchments, collectively referred to as the Maroondah catchments (163 km^2) (Figure 1). These form part of the total catchment area (1040 km^2) supplying the city of Melbourne, Australia (pop. 3·25 million). The largest of the Maroondah catchments contains a storage reservoir (1·9 km^2, 22 Gl). Hydrographs from the three largest catchments (145 km^2) are summed as one for this study, the two smallest catchments being unreliably gauged. At a daily level, differences in timing of flow from these catchments are unimportant.

The area is a long-term hydro-ecological research site featuring 18 small gauged experimental catchments within and around the water supply catchments (Figure 1). Research has focused on quantifying and predicting the long-term effects on water yield of land cover change resulting from wildfire or logging (see Langford, 1976; Kuczera, 1987; Vertessy *et al.*, 1994a).

The area exhibits steep, deeply incised terrain with laterally confined stream lines and relief of over 1000 m. Mean annual precipitation (MAP) varies markedly from around 1100 mm at lower elevations to over 2800 mm on the highest mountain top. Most of the precipitation falls as rain. In mid-winter, snow is common above 1000 m but rarely accumulates to more than 30 cm deep. The catchments are almost entirely forested and are dominated by pure stands of tall wet-sclerophyll ash-type forest, including 59% mountain ash (*Eucalyptus regnans*) and 7% other ash-type species. Dry-sclerophyll 'mixed species' eucalypt forest (22%) and cool temperate rainforest (5%) are the remaining major forest types. Only the ash forest has been studied in detail.

The area is characterized by very deep, permeable soils leading to deep, static water tables away from streams and shallow, dynamic and hydrologically significant water table response near streams. Shallow soils occur on certain ridge tops and north-facing slopes.

Stream flow originates from and responds to longitudinally extensive saturated areas that enlarge significantly during wet periods (Duncan and Langford, 1977; Finlayson and Wong, 1982). A high and relatively constant proportion of baseflow to total flow is maintained throughout the year. Mean annual runoff for the subject catchments is 743 mm (53-year record).

SPATIAL DISAGGREGATION OF THE STUDY AREA

The GRASS GIS, augmented by an array of purpose-written, GRASS-compliant GIS software, was used throughout.

A 25-metre digital elevation model (DEM) was interpolated from 20-metre digital contour data using spline interpolation (Mitasova and Hofierka, 1993) (Figure 1). From this, the upslope area per unit contour length for each cell was calculated using a multiple flow direction algorithm similar to that of Quinn *et al.* (1995). This formed the basis for the division of the catchment into hillslopes using the algorithm of Lammers and Band (1990). The hillslope size was chosen such that real hillslopes were delineated (a subjective judgement in this case). A total of 131 hillslopes resulted, with a mean area of 123 ha (1·23 km^2), a median of 89 ha, and a range of 0·75–491 ha.

The hillslopes were further disaggregated according to a wetness index. Technically, the form of the wetness index is constrained only by the requirement that it is static. However, it is desirable to choose an index that, when input to the distribution function, leads to accurate predictions of water table depth along a hillslope catena. The index so chosen was the topographic wetness index used with TOPMODEL, $\ln(a/\tan \beta)$ (where a is upslope area per unit contour length, and β is the terrain slope).

Each hillslope was divided into ESUs defined by sub-ranges of the wetness index, i.e. areas of similar wetness index were lumped together to form single ESUs. This resulted in 1848 ESUs in the entire study area (ranging from 0·6–168 ha, with a mean area of 8·7 ha and a median of 3·5 ha) and hence an average of 14 ESUs per hillslope.

INITIAL PARAMETERIZATION AND CALIBRATION

Because Macaque is a physically based model, the values of input parameters *and internal variables* must agree with the physical quantities that they represent. Macaque uses a large number of parameters and internal variables (the exact number depends upon what is considered to be 'the model' and what is considered to be external to the model). Most parameter values were derived from physical measurements (e.g. maximum leaf conductance). Others, particularly those applying to subsurface systems (e.g. hydraulic gradient within the saturated zone), were derived from calibration against various observations of internal variables as well as observed hydrographs. We have not used any optimization functions or automatic calibration procedures, but rather looked simultaneously at a range of variables whilst applying manual calibrations. Qualitatively, the criteria used to determine that the model was operating in a realistic manner are summarized as follows (quantitative results are given by Watson, 1997):

- predicted and observed hydrographs matched *approximately* at daily, weekly, monthly and yearly scales;
- all stream areas marked on a topographical map produced runoff;

- the runoff producing areas expanded in winter;

- the upslope and midslope areas of hillslopes did not produce runoff except in the highest rainfall areas;

- upslope water tables remained 10 metres or deeper below the soil surface (as observed using seismic survey);

- near-stream water tables remained within a few metres of the soil surface and varied by about one metre annually (as observed using piezometers);

- canopy and understorey transpiration remained near observed ranges (observed using heat-pulse probes);

- canopy conductance remained near observed ranges (observed data from Connor *et al.*, 1977);

- interception of precipitation matched observed totals and temporal patterns (observed data available from numerous studies); and

- predicted climatic variables, such as Vapour pressure deficit (VPD), and net radiation remained within locally observed ranges and displayed expected spatial patterns throughout the study area (numerous observed data available).

INITIAL MODEL APPLICATION

Results derived using a reference parameter set are given initially. With reference to the parameter mapping techniques described below, this parameter set used three-dimensional precipitation interpolation, species-based LAI mapping and air photograph interpreted topography. The results may be compared with results derived using the alternative parameter sets presented in the following sections. A three-year simulation period was used spanning the years 1982 to 1984. This covers a wide range of catchment conditions. The 1982–1983 summer was extremely dry, the 1984 winter was wetter than most winters. Whilst the model operates with a daily time step, weekly hydrographs are presented for clarity. Figure 2a shows predicted and observed hydrographs and Figure 3a maps the associated runoff source areas.

The predicted hydrograph follows the seasonal variation apparent in the observed hydrograph, perhaps underestimating the variability between the dry 1982 and wet 1984 winters. Predicted baseflow recession is quicker than is observed. Peaks resulting from storm flow match the observed peaks in character but tend to be overestimated in winter. This is likely to be associated with the assumption of constant hydraulic gradient within the water table near where it is exfiltrating as baseflow. Observations using piezometers show that the hydraulic gradient increases as the water table mounds during wet periods. Macaque, whilst able to predict mounding in the water table, does not as yet relate this to increases in hydraulic gradient. Thus, in order for baseflow to increase in wet periods, the saturated area must increase beyond realistic levels. This, in turn, leads to excessive storm flow, as shown in Figure 2a.

Predicted stream flow source areas (Figure 3a) are generally as expected. All mapped watercourses are represented as source areas, as are broad areas observed in the field as 'soaks'. The high precipitation areas in the south-east generate more stream flow from a greater proportion of each hillslope than other areas. A general failing of the model, however, is that stream flow is predicted to originate from a number of hillslopes known to be permanently unsaturated, such as the upper slopes of the south-west tip of the southern catchment. This is most likely related to the problem of excessive saturated areas noted above.

In the following sections, the model is run using alternative precipitation, LAI and topographic mapping schemes respectively, and the effects on the simulated hydrological response are examined.

PRECIPITATION MAPPING

Precipitation records are available from 73 past and present gauging sites within 15 km of the centre of the study area as shown in Figure 4, including eight gauges installed for the present study. This represents a

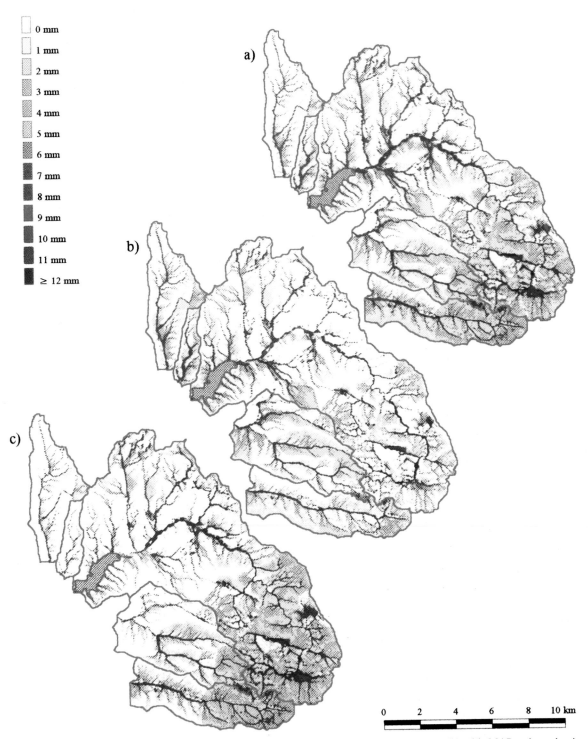

Figure 3. Mean daily stream flow source areas for 1984. (a) Using the reference parameter set; (b) with MAP estimated using one-dimensional linear instead of three-dimensional spline interpolation; and (c) with total LAI estimated from TM imagery instead of forest type

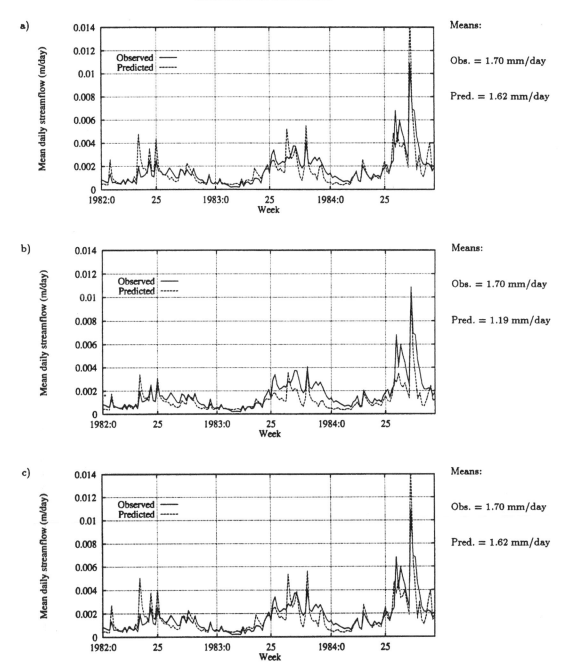

Figure 2. Predicted and observed weekly hydrographs aggregated from the three largest water supply catchments for the period 1982–1984. (a) Using the reference parameter set; (b) using one-dimensional linear instead of three-dimensional spline interpolation of precipitation; and (c) using total LAI estimated from TM satellite imagery instead of forest type

highly detailed data set ideal for the investigation of precipitation estimation techniques and their effects on hydrological model operation. Two such techniques are investigated here.

Precipitation distribution in mountainous regions is often estimated using a one-dimensional linear precipitation/elevation correlation. Given the common observation that precipitation increases with

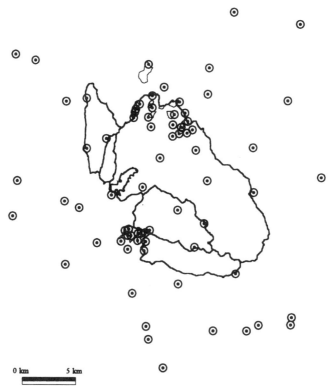

Figure 4. The 73 precipitation gauge sites within 15 km of the centre of the study area

elevation, this technique captures the basic pattern of precipitation distribution within a region. It is limited, however, where non-linear relationships are observed between precipitation and elevation (e.g. where precipitation approaches a limit above a certain elevation), or where horizontal spatial influences exist independently of vertical influences (e.g. proximity to the coast).

Figure 5a shows MAP estimated from one-dimensional linear regression of MMPI versus elevation at each of the 73 gauge sites. By definition, precipitation distribution exactly matches the DEM (Figure 1). The mean value for the three largest catchments is 1610 mm.

An alternative means of estimating precipitation distribution is to use some form of non-linear interpolation, and to use three spatial dimensions as independent variables. This allows more degrees of freedom in matching observed data and, providing a robust method is used, will more accurately reproduce observed spatial patterns. The cost of this accuracy and freedom is that data from a larger number of gauges is required in order that non-linearities and horizontal spatial patterns may be observed to begin with. A large amount of data are available for the Maroondah region. However, the investigations carried out here will also be of benefit in less intensely gauged regions by providing knowledge of precipitation patterns, which may be used to develop accurate, parsimonious estimation methods, and by quantifying the effect on model operation of choosing different interpolation schemes.

The ANUSPLIN package developed by Hutchinson (1995) was used to perform three-dimensional spline interpolation of MMPI values from the 73 gauge sites using elevation, easting and northing as independent variables. This resulted in the map of MAP shown in Figure 5b. The mean value for the three largest catchments is 1807 mm, 12% higher than if the one-dimensional linear technique were used. Note that this technique cannot be implemented with the two-dimensional spline interpolation routines available in many GIS (including GRASS) which use easting and northing as independent variables.

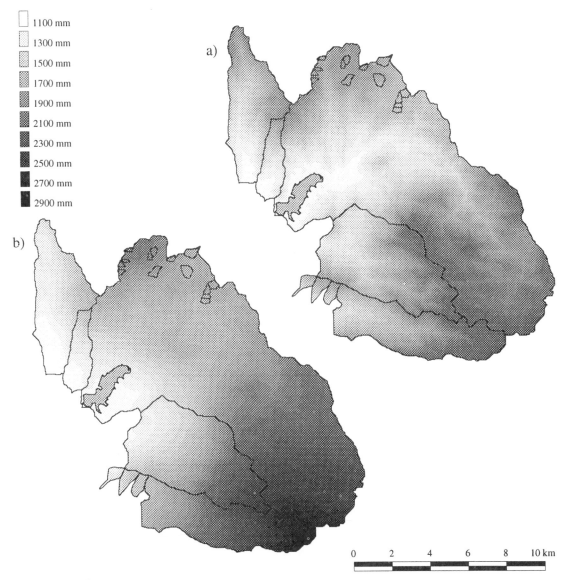

Figure 5. Mean annual precipitation estimated from (a) one-dimensional linear interpolation of MMPI values from 73 gauging sites and (b) three-dimensional spline interpolation of the same MMPI values

The general pattern shown by the two maps is similar; however, there are some distinct differences. The three-dimensional spline map predicts much higher precipitation in the south-east than the one-dimensional linear map. This is largely caused by high precipitation recorded by a single gauge atop Mount Donna Buang (see Figure 1). There are two reasons why the high precipitation in this area is not predicted using the one-dimensional linear technique. The first is that, because the technique uses linear regression, a single point will not affect the regression parameters very much. Indeed, the parameters will be more influenced by the clusters of gauges surrounding the experimental catchments, which are confined to two small regions of the study area. A more sophisticated one-dimensional interpolation scheme could remedy this. A second reason is that Mount Donna Buang is part of a mountain range that receives more precipitation relative to its elevation than mountain-top sites away from this range, such as those in the north of the experimental area.

Proximity to the mountain range represents a horizontal as well as a vertical influence on precipitation which can not be represented by any one-dimensional interpolation.

Using the otherwise unchanged reference parameter set, Macaque was run using each of the two MMPI maps. The hydrographs produced using three-dimensional spline interpolated MMPI and one-dimensional linear MMPI are shown in Figure 2a and 2b, respectively. The corresponding pair of maps of predicted stream flow source areas are shown in Figure 3a and 3b. The hydrograph for the one-dimensional linear map is lower and less peaky than that for the three-dimensional spline map. Mean daily flow is 27% lower using the one-dimensional linear map (or 36% higher, vice versa). In the south-east of the study area, there is a significant reduction in the proportion of hillslopes that generate stream flow under the one-dimensional linear map. These observations all relate to the lower precipitation predicted for the south east by the one-dimensional linear technique. The peakiness and excessive saturation of hillslopes predicted under the three-dimensional spline map were cited earlier as problems with the initial simulation so these observations favour the use of the more simple technique.

There is some uncertainty as to whether the more sophisticated mapping has resulted in improved prediction of stream flow because predictions made under the simpler one-dimensional linear technique exhibited more desirable features than those made using the 'improved', three-dimensional spline technique. The investigation has highlighted the sensitivity of precipitation mapping. It is important to note that the density of gauges available for this exercise was much greater than is usual. Nevertheless, the choice of analysis method greatly influenced the simulated response. Future measurements will need to constrain the prediction of other elements of the water balance to the point where more can be said about the true distribution of precipitation.

LEAF AREA INDEX MAPPING

LAI has been measured in the study area both as a result of destructive measurement, where trees are felled and their leaves are measured, and by ground-based remote sensing using a light sensor or 'LICOR' (e.g. Vertessy *et al.*, 1994b, 1995a). Additionally, canopy species and age have been mapped in detail by the Victorian Department of Natural Resources and Environment. These data have been used to produce two maps of total LAI (canopy plus understorey).

For the standard map, typical total LAI values were assigned to distinct forest types mapped on the basis of canopy species, as shown in Figure 6a. Spatial variability is dominated by the distribution of the three major forest types: mixed species (low LAI), ash (medium LAI) and rainforest (high LAI). The mean value for the three largest catchments is 3·78.

An alternative total LAI map was produced by combining LICOR measurements, destructive measurements and typical LAI values for areas of extreme LAI (both high and low) with satellite indices. Four Landsat TM images were obtained for this purpose. Each image was resampled to be geocoded in the Australian Map Grid (AMG) with 25-metre pixels, then systematically translated in a range of directions to align the image to the DEM by maximizing the correlation with DEM-derived theoretical shading images produced using the Minnaert (1941) method. The images exhibited strong topographical shading which was corrected separately for each of the seven bands in each image according to Colby (1991). Results from the four images were then averaged to produce a temporal 'mean' image comprised of seven bands. Maps of various satellite indices were computed from the mean image, including a normalized difference vegetation index (NDVI, see Barrett and Curtis, 1992) map (Band 4 − Band 3/Band 4 + Band 3) and an index using the same bands based on shade corrected imagery (corrected Band 4/Band 3).

These indices were plotted against LICOR observations of LAI in order to find a relationship predicting LAI from the satellite imagery. This was only partially successful because the range of measured LAI values was small. The corrected Band 4/corrected Band 3 image clearly distinguishes between areas of very low LAI (such as recently cleared areas) and areas of high LAI (such as 10-year old ash forest), so indices were sampled from these areas and added to the plot with typical LAI values for these forest types. A linear fit was

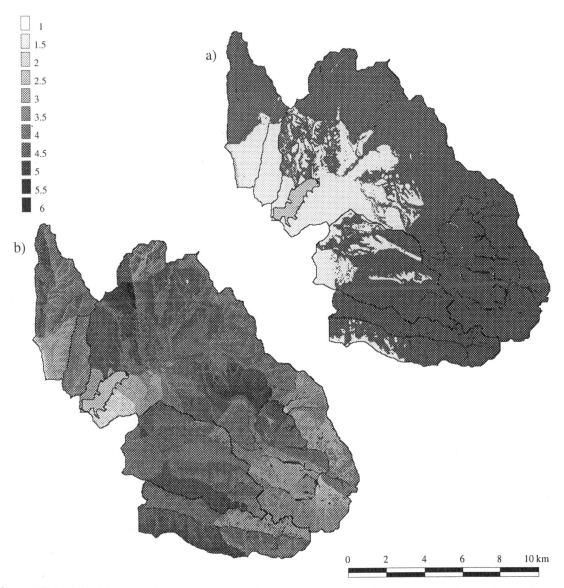

Figure 6. Total LAI, (a) Estimated by assigning generally accepted values to each forest type; and (b) estimated from TM satellite imagery

constructed from these data and applied to the corrected Band 4/corrected Band 3 image to give a map predicting total LAI, as shown in Figure 6b. The mean value for the three largest catchments is 3·77, which is very close to the average obtained from the species-based map (3·78).

The model requires that total LAI be split into canopy and understorey components. For the ash forest, canopy LAI was estimated from forest age using an equation constructed by Watson and Vertessy (1996) from destructive measurements. Constant canopy LAI values were assigned to other forest types. Understorey LAI was then calculated as the difference between total and canopy values.

The average LAIs predicted by the species-based and satellite-based maps differ by less than one per cent. Spatially, however, the two maps differ significantly. The species-based map shows large areas of uniform total LAI, broken only near the major boundary between ash-type and other forests (the latter occurring near the reservoir), and along the bands of rainforest bordering streams in high precipitation areas. The satellite-based map exhibits significantly more spatial variability. For example, a large area of low LAI is shown in the south-east corner, coinciding with a large area of old growth ash forest. There is also variability that appears to be aspect related, whereby numerous opposing north/south hillslopes have different LAIs. This may be real, or it may be some residual topographic shading effect resulting from inadequate shading correction, or it may be a combination of both. Insufficient data are available at this stage to clarify this, so there is uncertainty in both maps. The point here is that LAI is a vital parameter in any physical model of forest hydrology and the two maps represent equally viable, yet quite different alternatives for mapping this parameter.

Using the otherwise unchanged reference parameter set, Macaque was run using each of the two total LAI maps. The hydrographs produced using the species-based LAI map and the satellite-based LAI map are shown in Figure 2a and 2c, respectively. The corresponding pair of maps of predicted stream flow source areas are shown in Figure 3a and 3c. There is almost no difference in the shape of the two hydrographs, and no difference in the mean daily stream flow. This is because the mean LAI was virtually the same for the two. Spatial variations in LAI do not affect mean stream flow for the study area. This suggests that evapotranspiration as predicted by Macaque is linearly proportional to LAI and probably not limited by spatially variable influences such as soil moisture stress. Whilst this observation satisfies our perceptual model of ash forest behaviour, it may not be appropriate for drier, water-limited forests. There are small differences in the maps of stream flow source areas, with the satellite based map reducing excessive hillslope saturation in drier areas. This indicates that the satellite-based map may be a better predictor of LAI, but further testing against observed data is needed before this assessment can be confirmed.

TOPOGRAPHY MAPPING

Topographical data are available from two sources. The first is digital contour data provided by the Victorian Division of Survey of Mapping and derived from air photographic interpretation (API). This is available for the entire catchment area in sufficient detail to construct a 25-metre gridded digital elevation model (DEM). The second source is a detailed ground survey of the Ettercon 3 experimental catchment (15 ha) (Figure 1) conducted as part of the current study, from which detailed DEMs have been produced. Watson *et al.* (1996) have shown that the API data describe a smoother terrain than reality, which is better expressed by the ground-based data. The API data are, however, the only data available for the *entire* study area and must be used as such.

The two DEMs were used to construct two separate maps of the $\ln(a/\tan\beta)$ wetness index, which were discretized into ESUs to provide two alternative topographic parameterization and spatial disaggregation schemes for Ettercon 3. Figure 7 shows the disaggregation into ESUs of the wetness index. The smooth representation of topography by the API DEM is reflected in Figure 7a which shows only a limited convergence of high wetness areas along the stream. Figure 7b, on the other hand, shows a concentrated gully with high wetness reflecting the incised topography more accurately represented by the ground surveyed DEM. This also more accurately reflects the shape of the ground surveyed saturated zone (during a relatively dry period) which is overlaid on both maps.

Macaque was run on each representation of Ettercon 3 with the otherwise unchanged reference parameter set. The resulting hydrographs are shown in Figure 8, and the predicted stream flow source areas are shown in Figure 9. The two predicted hydrographs are almost identical and are reasonably accurate in both shape and mean value. Neither hydrograph predicts the effects of the drought in the summer of 1982–1983 very well, and both hydrographs are too variable during the high flow period at the end of 1984. The fact that such a marked difference in the two DEMs used to produce these hydrographs does not induce a difference in the hydrographs is perhaps surprising. The maps of runoff source area, however, are quite different. Field observations indicate that Figure 9a is an unrealistic simulation of the runoff producing area. The smoothing of terrain apparent in the API-derived map clearly spreads out the saturated area too much. Despite this

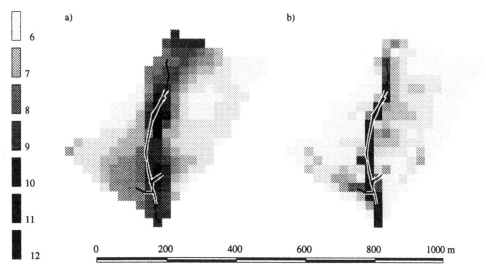

Figure 7. ESUs calculated from the $a/\tan\beta$ wetness index for the Ettercon 3 experimental catchment calculated from (a) the API DEM; and (b) the ground surveyed DEM. Both maps are overlaid with ground surveyed stream lines and saturated area boundaries (during a relatively dry period)

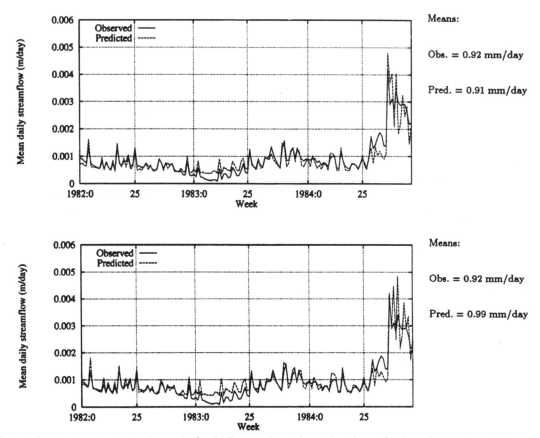

Figure 8. Predicted and observed weekly hydrographs for the Ettercon 3 experimental catchment for the period 1982–1984; (a) Using the API DEM; and (b) using the ground surveyed DEM

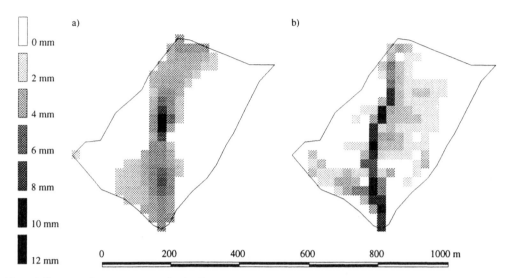

Figure 9. Mean daily stream flow source areas for the Ettercon 3 experimental catchment in 1984 using (a) the API DEM and (b) the ground surveyed DEM

spreading of the saturated area, it is clear from the hydrograph response that the effective evapotranspiration is unaffected by this spatial smoothing. It appears that whilst the maps of wetness index differ, the statistical representation of topography that the two DEMs effect within the distribution function is almost the same from a hydrological point of view. This indicates that there would be little benefit from obtaining more detailed topographic data over the whole catchment because the structure of the distribution function model is unable to exploit the greater detail.

DISCUSSION

In applying LSSMs, the user is faced with a myriad of questions regarding the way in which spatial information is to be used. In the case of data such as precipitation, point information must be interpolated whilst variables such as LAI must be derived from surrogate measures such as forest type or electromagnetic reflectance. The way in which the modeller chooses to use the information has important implications for model response. There is always great uncertainty regarding the true nature of spatially variable inputs and it is not always the case that apparently more sophisticated analytical approaches yield more certain results.

The three-dimensional interpolation of precipitation was strongly influenced by data from a particular gauge so it was impossible to say whether the resulting rainfall distribution was a more realistic representation or simply an artefact of an unrepresentative gauge.

Similar uncertainty existed in the estimation of LAI, a parameter to which models of forested areas are highly sensitive. Fortunately, the mean LAI estimates by the two methods were similar so overall runoff was similar, but there were large spatial differences in LAI representation. The spatial variability of the satellite-derived data gave a more realistic looking spatial runoff response but there are no direct data to confirm this observation. Ultimately, the user must simply accept that by choosing a particular method of representation, a certain spatial variability of response is 'locked in'.

It must also be accepted that different model structures can be limited in the extent to which they can exploit more detailed information. In the example given above, more detailed topographic data of a small subcatchment gave a clearly more realistic estimate of runoff producing area yet, even at this small scale, the impact on the hydrograph was not detectable. This was because the representation of processes (especially evapotranspiration) within the model structure makes little distinction between saturated and almost saturated soil. There would be no point in trying to obtain such detailed topographic data for the entire area.

Model users must consider carefully whether the model they are using is able to exploit more detailed information.

Ultimately, the choice of spatial detail depends on the user's ability to assess whether the simulated response is a better representation of reality. This can be both a quantitative and a qualitative assessment and should consider not only whether the specific variable of interest is better represented but also whether the simulated response of interest is affected. If the former is true but the latter is not, there may still be no value in greater spatial detail since the model structure is unable to use the higher level of information.

ACKNOWLEDGEMENTS

We thank: Richard Silberstein, Larry Band, Scott Mackay, Richard Lammers, Richard Fernandes, Rob Lamb and Keith Beven for ideas and advice on modelling; Richard Campbell, Sharon O'Sullivan, Sharon Davis, Campbell Pfeiffer, Judi Buckmaster and Andrew Western (all from Melbourne or Monash Universities, Melbourne, Australia) for providing unpublished data on water tables, soil hydraulic conductivity, transpiration, LAI and radiation; Petina Pert and Jill Smith of the Victorian Department of Natural Resources and Environment for supplying digital forest type data; and two reviewers for their helpful comments.

REFERENCES

Band, L. E., Patterson, P., Nemani, R., and Running, S. 1993. 'Forest ecosystem processes at the watershed scale: incorporating hillslope hydrology'. *Agric. For. Meteorol.*, **63**, 93–126.

Barrett, E. C. and Curtis, L. F. 1992. *Introduction to Environmental Remote Sensing*, (3rd edn). Chapman & Hall, London. 426 pp.

Beven, K. J., Lamb, R., Quinn, P., Romanowicz, R., and Freer, J. 1994. 'TOPMODEL', in Singh, V.P. (ed.), *Computer Models of Watershed Hydrology*. Water Resources Publications, Highlands Ranch, Colorado. pp. 627–668.

Colby, J. D. 1991. 'Topographic normalisation in rugged terrain', *Photogramm. Engng Remote Sens.*, **57**, 531–537.

Connor, D. J., Legge, N. J., and Turner, N. C. 1977. 'Water relations of Mountain Ash (*Eucalyptus regnans* F. Muell.) forests', *Aust. J. Plant Physiol.*, **4**, 753–762.

Duncan, H. P. and Langford, K. J. 1977. 'Stream flow characteristics', in Langford, K. J. and O'Shaughnessy, P. J. (eds), *First Progress Report — North Maroondah*, Rep. No. MMBW-W-0005. Water Supply Catchment Hydrology Research, Melbourne and Metropolitan Board of Works. pp. 215–229.

Finlayson, B. L. and Wong, N. R. 1982. 'Storm runoff and water quality in an undisturbed forested catchment in Victoria', *Aust. For. Res.*, **12**, 303–315.

Hatton, T., Dyce, P., Zhang, L., and Dawes, W. 1995. 'WAVES — an ecohydrological model of the surface energy and water balance: sensitivity analysis'. *Tech. Memo. 95.2*. CSIRO, Institute of Natural Resources and Environment, Division of Water Resources, Canberra.

Hutchinson, M. F. 1995. 'Interpolating mean rainfall using thin plate splines'. *Int. J. Geograph. Inform. Syst.*, **9**, 385–403.

Kuczera, G. A. 1987. 'Prediction of water yield reductions following a bushfire in ash-mixed species eucalypt forest', *J. Hydrol.*, **94**, 215–236.

Lammers, R. B. and Band, L. E. 1990. 'Automating object representation of drainage basins', *Comp. Geosci.*, **16**, 787–810.

Langford, K. J. 1976. 'Change in yield of water following a bushfire in a forest of Eucalyptus regnans', *J. Hydrol.*, **29**, 87–114.

Minnaert, M. 1941. 'The reciprocity principle in lunar photometry'. *The Astrophys. J.*, **93**, 403–410.

Mitasova, H. and Hofierka, L. 1993. 'Interpolation by regularized spline with tension: II. Application to terrain modeling and surface geometry analysis'. *Math. Geol.*, **25**, 657–669.

Quinn, P. F., Beven, K. J., and Lamb, R. 1995. 'The ln (a/tan β) index: how to calculate it and how to use it within the TOPMODEL framework', *Hydrol. Process.*, **9**, 161–182.

Silberstein, R. P. and Sivapalan, M. 1995. 'Estimation of terrestrial water and energy balances over heterogeneous catchments', *Hydrol. Process.*, **9**, 613–630.

Vertessy, R. A., Hatton, T. J., O'Shaughnessy, P. J., Jayasuriya, M. D. A. 1993. 'Predicting water yield from a mountain ash forest catchment using a terrain analysis based catchment model'. *J. Hydrol.*, **150**, 665–700.

Vertessy, R. A., Benyon, R., and Haydon, S. 1994a. 'Melbourne's forest catchments: Effect of age on water yield', *Water, J. Aust. Wat. Wastewat. Ass.*, **21**, 17–20.

Vertessy, R. A., Benyon, R. G., O'Sullivan, S. K., and Gribben, P. R. 1994b. 'Leaf area and tree water use in a 15 year old mountain ash forest, Central Highlands, Victoria', *Report No. 94/3*. Cooperative Research Centre for Catchment Hydrology, Monash University, Clayton, Victoria, Australia. 31 pp.

Vertessy, R. A., Benyon, R. G., O'Sullivan, S. K., and Gribben, P. R. 1995a. 'Relationships between stem diameter, sapwood area, leaf area and transpiration in a young mountain ash forest', *Tree Physiol.*, **15**, 559–567.

Vertessy, R. A., Hatton, T. J., Benyon, R. J., and Dawes, W. R. 1995b. 'Long term growth and water balance predictions for a mountain ash (*Eucalyptus regnans*) forest catchment subject to clearfelling and regeneration', *Tree Physiol.*, **16**, 221–232.

Watson, F. G. R. 1997. 'Physically based prediction of water yield from large forested catchments: integrated hydro-ecological modelling of the Maroondah Catchments, Victoria', *PhD Thesis*. Department of Civil and Environmental Engineering, Melbourne University, Melbourne.

Watson, F. G. R. and Vertessy, R. A. 1996. 'Estimating leaf area index from stem diameter measurements in Mountain Ash forest', *Report 96/7*. Cooperative Research Centre for Catchment Hydrology, Monash University, Clayton, Victoria, Australia. 102 pp.

Watson, F. G. R., Vertessy, R. A., and Band, L. E. 1996. 'Distributed parameterization of a large scale water balance model for an Australian forested region', *HydroGIS 96: Application of Geographic Information Systems in Hydrology and Water Resources Management, Proceedings of the Vienna Conference, April, 1996, IAHS Publ.*, **235**, 157–166.

Watson, F. G. R., Vertessy, R. A., and Grayson, R. B. 1997. 'Large scale, long term, physically based prediction of water yield in forested catchments', pages 397–402 of proceedings, *International Congress on Modelling and Simulation (MODSIM '97), Hobart, Tasmania, 8–11 December, 1997*.

6

APPLICATION OF A GIS-BASED DISTRIBUTED HYDROLOGY MODEL FOR PREDICTION OF FOREST HARVEST EFFECTS ON PEAK STREAM FLOW IN THE PACIFIC NORTHWEST

PASCAL STORCK,[1]* LAURA BOWLING,[1] PAUL WETHERBEE[2] AND DENNIS LETTENMAIER[1]

[1]*Department of Civil Engineering, Box 352700, University of Washington, Seattle, WA 98195, USA*
[2]*HDR Alaska, Inc., Anchorage, AK 99503-2639, USA*

ABSTRACT

Spatially distributed rainfall–runoff models, made feasible by the widespread availability of land surface characteristics data (especially digital topography), and the evolution of high power desktop workstations, are particularly useful for assessment of the hydrological effects of land surface change. Three examples are provided of the use of the Distributed Hydrology-Soil–Vegetation Model (DHSVM) to assess the hydrological effects of logging in the Pacific Northwest. DHSVM provides a dynamic representation of the spatial distribution of soil moisture, snow cover, evapotranspiration and runoff production, at the scale of digital topographic data (typically 30–100 m). Among the hydrological concerns that have been raised related to forest harvest in the Pacific Northwest are increases in flood peaks owing to enhanced rain-on-snow and spring radiation melt response, and the effects of forest roads. The first example is for two rain-on-snow floods in the North Fork Snoqualmie River during November 1990 and December 1989. Predicted maximum vegetation sensitivities (the difference between predicted peaks for all mature vegetation compared with all clear-cut) showed a 31% increase in the peak runoff for the 1989 event and a 10% increase for the larger 1990 event. The main reason for the difference in response can be traced to less antecedent low elevation snow during the 1990 event. The second example is spring snowmelt runoff for the Little Naches River, Washington, which drains the east slopes of the Washington Cascades. Analysis of spring snowmelt peak runoff during May 1993 and April 1994 showed that, for current vegetation relative to all mature vegetation, increases in peak spring stream flow of only about 3% should have occurred over the entire basin. However, much larger increases (up to 30%) would occur for a maximum possible harvest scenario, and in a small headwaters catchment, whose higher elevation leads to greater snow coverage (and, hence, sensitivity to vegetation change) during the period of maximum runoff. The third example, Hard and Ware Creeks, Washington, illustrates the effects of forest roads in two heavily logged small catchments on the western slopes of the Cascades. Use of DHSVM's road runoff algorithm shows increases in peak runoff for the five largest events in 1992 (average observed stream flow of $2 \cdot 1$ m^3 s^{-1}) averaging 17·4% for Hard Creek and 16·2% for Ware Creek, with a maximum percentage increase (for the largest event, in Hard Creek) of 27%.

KEY WORDS   catchment modelling; anthropogenic effects; floods

* Correspondence to: Pascal Storck, Department of Civil Engineering, Box 352700, University of Washington, Seattle, WA 98195, USA.

Contract grant sponsors: NASA; State of Washington Department of Natural Resources; National Council of the Paper Indsutry for Air and Stream Improvement.
Contract grant numbers: FY96-088 (Washington); CW-14 (Paper Industry).

INTRODUCTION

Hydrological simulation (sometimes termed rainfall–runoff) modelling began in the 1950s and 1960s with the advent of the digital computer. The purpose was to predict stream flow, given observed precipitation (and perhaps other meteorological variables), at time-scales that were short compared with catchment storm response times. Among the various applications of hydrological simulation models are stream flow forecasting, design and planning (e.g. for flood protection), and extension of stream flow records. The first models, of which the Stanford Watershed Model of Crawford and Linsley (1966) is perhaps the best-known example, were spatially lumped, meaning that the models represented the effective response of an entire catchment, without attempting to characterize the spatial variability of the response explicitly. Precipitation forcings were usually represented as mean areal precipitation, and typically were obtained by spatial averaging of gauge observations.

Spatially lumped stream flow simulation models are still widely used; for instance, the backbone of the National Weather Service's River Forecast System is the Sacramento Soil Moisture Accounting Model (Burnash *et al.*, 1973) which is an outgrowth of the Stanford Watershed Model. None the less, a critical shortcoming of spatially lumped (often termed conceptual) hydrological simulation models is their inability to represent the spatial variability of hydrological processes and catchment parameters (Moore *et al.*, 1991). Furthermore, the effective parameters used by these models are not easily related to physically observable surface characteristics, which makes parameter estimation a time consuming, and often scientifically unsatisfying experience.

The last 10 years have seen the development of a second generation of hydrological simulation models, which are able to represent explicitly the spatial variability of some, if not most, of the important land surface characteristics (see Moore *et al.*, 1991 for a review). Two factors have made these recent developments possible. First, digital topographic and other land surface data are now widely available. For instance, three arc-second (roughly 90 m N–S) topographic data for the continental US can be obtained for a nominal charge, and similar data at one arc-second resolution are available for much of the country. The availability of digital topographic data at these resolutions elsewhere in the world varies, but coarser resolution topographic data are available globally (e.g. the recently released Digital Chart of the World includes 500 m resolution digital topographic data). Likewise, spatially distributed data sets for other surface variables important to hydrologists, such as soils data (e.g. from the US Natural Resources Conservation Service's STATSGO and SSURGO data sets) and vegetation data, are now readily available. Certainly, the accuracy and reliability of spatially distributed land data sets varies. None the less, with respect to land surface data, hydrologists now find themselves in a data-rich situation, which is in direct contrast to the data-poor situation that often exists for hydrological variables (such as, for instance, stream flow, precipitation, soil moisture, and snow).

The second factor that has spurred the development of spatially distributed hydrological modelling is the explosion in desktop computing power. Most hydrological modelling applications that were run on large mainframe computers 10 years ago are now performed on UNIX workstations. A second migration to PC platforms from UNIX workstations is underway, as the capabilities of personal computers now overlap those of low- to mid-range UNIX workstations. The growth of desktop computing has made possible the development of software tools, such as GIS and database management systems, that are capable of processing image-type data, such as land surface characteristics, on desktop computers. These software tools have, to a large extent, replaced the site-specific software previously used in many mainframe applications, which has had the effect of greatly enhancing the transportability of models and model applications.

Beyond their ability to relate model parameters directly to physically observable land surface characteristics, spatially distributed hydrological models have important applications to the interpretation and prediction of the effects of land use change. Because the models incorporate image-type representations of land surface conditions, prediction of the effects of changes in such conditions (including, especially, vegetation) are straightforward. In this paper, we describe the approach followed in using the Distributed

Hydrology–Soil–Vegetation Model (DHSVM; Wigmosta *et al.*, 1994) to predict the effects of forest harvest on peak stream flows in the Pacific Northwest. We present a brief description of DHSVM, followed by a more detailed discussion of model input requirements and preprocessing using GIS. Calibration issues are then discussed in the context of forest harvest assessment. Model output is presented in the context of three applications highlighting those data that provide insight for forest harvest effects.

BACKGROUND

In the mountainous areas of the north-western US, what amounts to a large-scale land use experiment has taken place in the last half-century, with the removal of much of the old growth, primarily coniferous, forest, by clear-cut logging, and its replacement with younger stands of more uniform age, and reduced species diversity. Land use change on such a vast scale is not unprecedented. During the 1800s, native grasslands and forests of the central US were converted to agricultural use. However, most of the central US land use change occurred prior to the advent of modern hydrological records; therefore, it is difficult to document the nature and magnitude of the associated hydrological process changes. On the other hand, most of the removal of old-growth forests from upland areas in the north-western US has occurred since World War II, during a period when the USGS stream gauge network had reached its maximum extent.

Anderson and Hobba (1959) were apparently the first to argue that the size of peak flows in western Oregon had been increased by logging, but the issue was not addressed in much detail until the work of Harr in the 1980s (see, for example Christner and Harr, 1982; Harr, 1981, 1986; Berris and Harr, 1987). Harr's plot-scale work demonstrated fairly conclusively that peak flows, especially for modest flood peaks caused by rain-on-snow events, could be significantly increased by clear-cutting, because of greater accumulations of snow in cleared areas prior to rain-on-snow events (as a consequence of reduced melt of snow intercepted by the canopy), and because of enhanced latent and sensible heat transfer in harvested areas, resulting from increased surface wind. Others have argued that forest roads are a more important contribution to post-logging runoff changes. For instance, Jones and Grant (1996) found that installation of a logging road network led to increases in stream flows comparable to those expected from clear-cutting, for small storm events.

In recent years, attention to floods during rain-on-snow events (and consequently forest harvesting) has increased. During the winters of 1986, 1990 and 1995–1996, the major rivers draining the western Cascade mountains of Washington and Oregon experienced flooding events with estimated return periods that exceeded 100 years in some cases. Parts of south-western Washington experienced two severe flood events during 1995–1996; one in November and another in February. These events have lead to a widespread perception that the severity and frequency of extreme floods in the Pacific Northwest has increased. Although much of this perception may be a result of climate variability, which has brought an end to the mild, relatively dry winters of the 1970s and 1980s, attention has nevertheless been focused on land use changes related to forest harvesting as a possible causal agent. This connection is not surprising given the results of the plot scale studies mentioned previously.

None the less, evaluation of the catchment-scale effects of logging has been more problematic. First, the magnitude of the enhanced rain-on-snow effect at the catchment scale is critically dependent on the extent of the rain-on-snow zone in any particular storm, and on the antecedent snow accumulation, and its nature. Secondly, the relative importance of any enhancement that might take place depends on the relative contribution of enhanced snowmelt (for very large storms, much of the snow may melt during the storm whether or not the rate is enhanced, hence the effect of land use change tends to be smaller). Finally, the spatial distribution of precipitation, and its interaction with the vegetation cover, plays a strong role in determining the magnitude of any enhancement. For this reason, catchment-scale assessments of changes in flood peaks that might be attributed to changed land use have had mixed results. Jones and Grant (1996) performed paired catchment comparisons of the peak stream flow series for three small catchments (60–101 ha) in the H. J. Andrews Experimental Watershed in western Oregon, as well as three sets of paired watersheds (typical areas several hundred km^2) with different harvest histories, and found that stream flow peaks in the most harvested catchments had generally increased relative to those in the catchments with less harvest. On the other hand, Connelly *et al.* (1993) evaluated trends in the annual peak stream flow series for 10 western Washington streams with gauge records extending back to the 1930s, and found significant changes in only one record that could reasonably be attributed to land use changes. They also found no significant change for the catchment with the largest apparent land use change.

Consequently, there is a strong need for hydrological modelling tools that can be used to assess the likely effects of land use changes on catchment-scale (e.g. flood) response. Watershed analysis, which is essentially a procedure for quantifying the ecological effects of forest practices prior to approval of harvest plans, has been adopted by both state and federal agencies in the Pacific Northwest as a management tool (Montgomery *et al.*, 1995). Watershed analysis is intended to develop forest management plans tailored for each watershed by defining areas most sensitive to the land use changes associated with forest harvesting. However, to identify these critical areas requires detailed information on the distribution of attributes over the watershed that can affect antecedent soil moisture, as well as snow accumulation and ablation. These attributes include topography, aspect, slope, soil properties, vegetation type and age, as well as the distribution of stream channels and logging roads. Distributed hydrological models are ideally suited to the needs of watershed analysis.

THE DISTRIBUTED SOIL–HYDROLOGY–VEGETATION MODEL (DHSVM)

DHSVM, as originally developed by Wigmosta *et al.* (1994) and extended for use in maritime mountainous watersheds (Storck *et al.*, 1995), provides a dynamic (one day or shorter time-step) representation of the spatial distribution of soil moisture, snow cover, evapotranspiration and runoff production. It consists of a two-layer canopy representation for evapotranspiration, a two-layer energy balance model for snow accumulation and melt, a multilayer unsaturated soil model and a saturated subsurface flow model. Model inputs are near-surface meteorology (precipitation, temperature, wind, humidity) and incoming short- and long-wave radiation. Digital elevation data are used to model topographic controls on incoming short-wave radiation, precipitation, air temperature and downslope water movement. Surface land cover and soil properties are assigned to each digital elevation model (DEM) grid cell or pixel. The DEM resolution is arbitrary, but the land surface is usually represented with pixels of dimension less than 100 m by 100 m.

In each model pixel, the land surface may be composed of overstorey vegetation, understorey vegetation and soil. The overstorey may cover all, or a prescribed fraction, of the land surface. The understorey, if present, covers the entire ground surface. The model allows land surface representations ranging from a closed two-storey forest, to sparse low-lying natural vegetation or crops, to bare soil. Meteorological conditions (precipitation, air temperature, solar radiation, wind speed, vapour pressure) are prescribed at a specified reference height well above the overstorey.

Solar radiation and wind speed are attenuated through the two canopies. If snow is present, it is assumed to cover the understorey and thus affects radiation transfer and the wind profiles via increased albedo and decreased surface roughness. Temperature and relative humidity are not adjusted through the canopy.

An independent one-dimensional (vertical) water balance is calculated for each pixel (Wigmosta *et al.*, 1994). Evaporation of intercepted water from the surfaces of wet vegetation is assumed to occur at the potential rate. Transpiration from dry vegetative surfaces is calculated using a Penman–Monteith approach. The model follows Entekhabi and Eagleson (1989) in using a soil physics-based approach to calculate soil evaporation.

Precipitation occurring below a threshold temperature is assumed to be snow. Snow interception by the overstorey is calculated as a function of the leaf area index (LAI) and is adjusted downwards for windy or cold conditions (Schmidt and Troendle, 1992). Intercepted snow can be removed from the canopy through snowmelt, sublimation and mass release. Melt of intercepted snow is calculated based on a single-layer energy balance approach. Mass release occurs if sufficient meltwater is generated during an individual time step such that the snow slides off the canopy (Bunnell *et al.*, 1985; Calder, 1990). Drip from the canopy is added to the ground snowpack (if present) as rain, while the cold content of any mass release or unintercepted snow is added directly to the ground snowpack.

Ground snow accumulation and melt are simulated using a two-layer energy balance model at the snow surface, similar to that described by Anderson (1968). The model accounts for the energy advected by rain, throughfall or drip, as well as net radiation and sensible and latent heat. Bulk transfer coefficients for turbulent exchange are calculated based on the aerodynamic resistance from the snow surface to the calculated two-metre wind and adjusted for atmospheric stability.

Unsaturated moisture movement through the soil layers is calculated using Darcy's law. This downward moisture flux recharges the grid cell water table. Each DEM grid cell in turn exchanges saturated subsurface flow with its eight adjacent neighbours according to topographic slope. This method allows a transient, three-dimensional representation of saturated subsurface flow. Return flow and saturation overland flow are generated in locations where grid cell water tables intersect the ground surface.

Runoff generation can be routed towards the watershed outlet by a distributed velocity surface routing algorithm (Maidment *et al.*, 1996). Surface runoff generated at a given pixel moves directly towards the watershed outlet based on a unit hydrograph that can include both a linear translation and a storage component. The travel time for each pixel is calculated based on its flow path and the velocity vectors along that flow path. The local pixel velocity is determined by the upstream drainage area and the local slope. Although this method is computationally simple, it does not allow for downstream reinfiltration.

An alternative method for surface runoff and open channel routing uses explicit information on the location of stream channels and road networks (Perkins *et al.*, 1996; Bowling and Lettenmaier, 1997). These two networks are imposed on the digital elevation model of the watershed topography as GIS coverages of vectors mapped to specific pixels. The fraction of each pixel covered by a road or stream is prescribed along with the depth of the road cut and/or channel incision. Precipitation is directly intercepted by these networks while subsurface flow is discharged directly into these networks. To determine the flux across the channel cut, we assume that the channel bottom is centred in the pixel while the groundwater gradient varies linearly from the channel bottom to its maximum height at the pixel boundary. Once in these networks, runoff is routed though the combined road and channel network using a Muskingum–Cunge scheme. Surface water not in the channel or road networks is modelled as overland flow and can reinfiltrate into neighbouring model pixels or be intercepted by the road or stream networks.

DHSVM INPUT, OUTPUT AND CALIBRATION

While not limited to any particular GIS product, DHSVM is ideally suited for integration into a GIS based analysis. The model requires detailed information on the meteorology and land surface for a given watershed, much of which is readily available in modern GIS formats. DHSVM input files fall into two general categories: time-series data and land surface maps. The majority of the meteorology is represented by time series data at the model's computational time-step (typically in the range of one hour to one day). Land surface data are represented by constant grid- or pixel-based maps. A few critical meteorological variables can be represented as a time series of maps at a frequency less than or equal to the model time-step, for example, monthly average clear-sky, incoming short-wave radiation for each model time-step. Given these needs, GIS software greatly facilities input data management. Furthermore, since DHSVM explicitly represents the landscape at fine spatial and temporal resolutions, the potential volume of output data makes GIS the preferred choice for synthesis and analysis.

Figure 1 shows a schematic representation of DHSVM, its inputs and outputs. Raw data describing the basin's topography and land surface are preprocessed with GIS software to produce the required DHSVM input files in either a binary or ASCII format. Data for the meteorological time series are used to force the model during the hydrological simulation and are also used to adjust the meteorological time series maps accordingly. Output data are stored for later analysis.

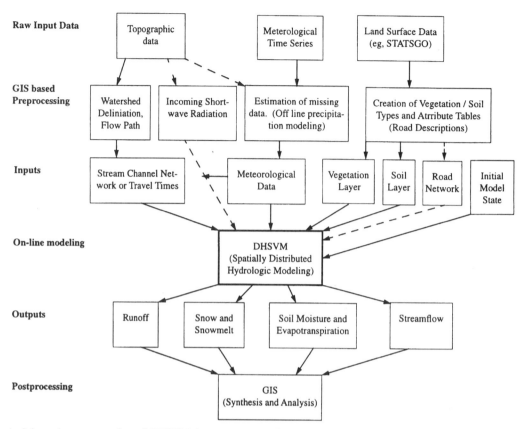

Figure 1. Schematic representation of DHSVM inputs, preprocessing requirements, output and interaction with GIS software. DHSVM options are indicated with dashed arrows

The raw input data for DHSVM falls into three main categories: (1) topography; (2) time-series meteorological data; and (3) land surface characterization. Each is described briefly below.

Topography

Digital elevation data are required to describe the topography of the watershed. The DEM is the fundamental foundation on which DHSVM and all of its distributed parameters, are based. For the three applications in this paper, topographic data were obtained from US Geological Survey 1 or 3 arc-second data and resampled to the square pixel dimension required using the ARC/INFO RESAMPLE command.

As shown in Figure 1, many other DHSVM inputs are calculated directly from the DEM. Watershed boundaries are determined with GIS-based utilities that track flow directions through the watershed such as the ARC/INFO FLOWDIRECTION and WATERSHED routines. Similar routines are used to define the local slope and accumulated drainage area for each pixel. Slope, accumulation and flow path information are used to calculate the travel time from a given pixel to the basin outlet for use with the unit hydrograph routing method. The stream channel network is imposed on the DEM from the flow accumulation map by specifying a minimum number of contributing pixels. Ideally, the synthesized network is verified against field observations. Once the stream channel network has been located, GIS routines are used to segregate the network into distinct reaches based on stream order, and assign attributes such as channel width, depth and roughness to each reach. A stream network input file generated by ARC/INFO contains the 'from' and 'to' nodes of each reach (e.g. flow out of reach 25 and 34 flow into reach 14).

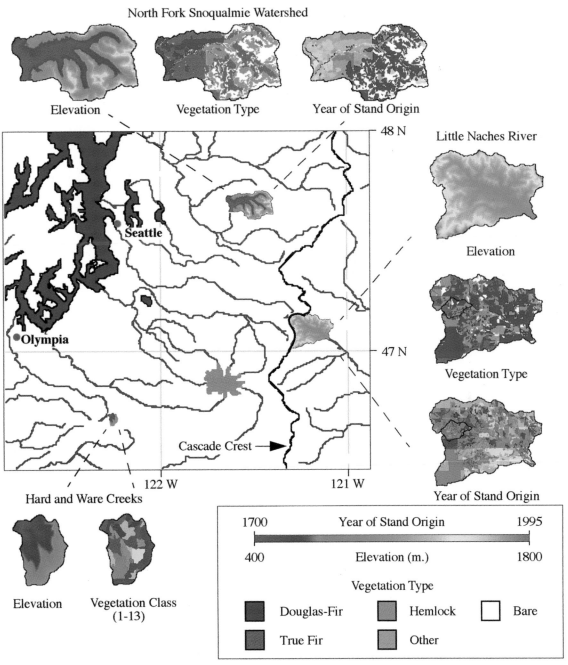

Figure 2. Location, elevation and vegetation species and age for the North Fork Snoqualmie, Little Naches and Hard/Ware creek watersheds

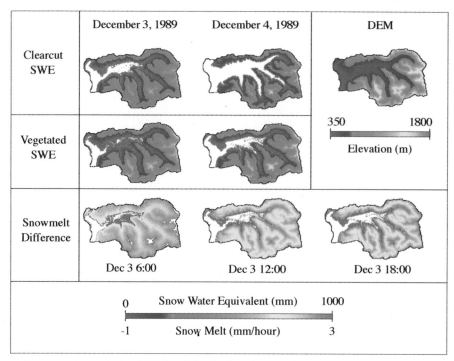

Figure 4. Change in snow water equivalent (SWE) and distribution of snowmelt owing to forest harvesting over the North Fork Snoqualmie basin. Uncoloured areas indicate zero in SWE images and no melt in either scenario for the difference images

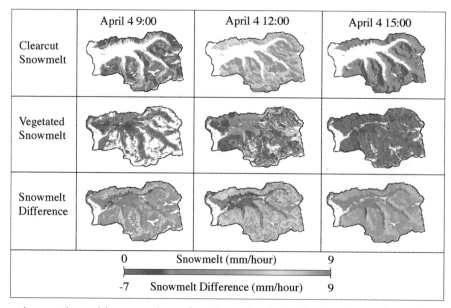

Figure 5. Snowmelt generation and increase owing to forest harvesting during the April 1990 spring melt event over the North Fork Snoqualmie basin. Uncoloured areas indicated zero in absolute snowmelt images and no melt in either scenario for the difference images

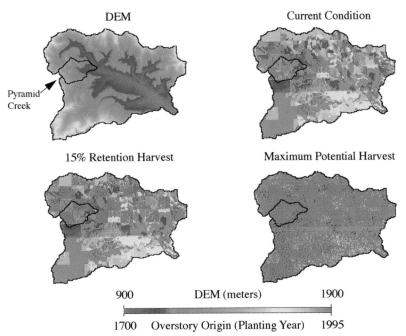

DEM   Current Condition

Pyramid
Creek

15% Retention Harvest   Maximum Potential Harvest

900   DEM (meters)   1900

1700 Overstory Origin (Planting Year) 1995

Figure 6. Little Naches river DEM and overstorey year of origin for current condition and two harvest prescriptions. The location of Pyramid Creek is also shown

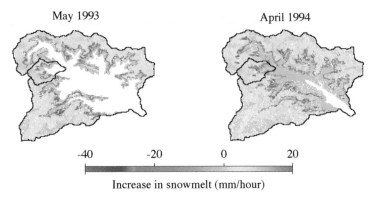

May 1993   April 1994

-40 -20 0 20

Increase in snowmelt (mm/hour)

Figure 7. Increase in snowmelt caused by maximum potential harvest compared with complete mature vegetation over the Little Naches River. Uncoloured areas indicate pixels with no snowmelt or no change in snowmelt

Meteorology (maps and time series data)

Based on the DEM, topographic controls on key meteorological variables can be determined. For example, the distribution of clear-sky, incoming short-wave radiation over the basin can be calculated for the monthly solar mid-points (e.g. 11 June, 17 July) for each day of the month for each time step (e.g. 6 to 9 a.m.). In our applications, we have used the model of Dubayah *et al.* (1990) as coded in the image processing workbench (IPW) (Frew, 1990; Longley, 1992) to generate maps of incoming direct beam and diffuse short-wave radiation. This allows us to include the effect of terrain reflectance and shadowing, which can play an important role in determining location and rate of snowmelt, in our analysis of forest harvesting effects (see, for example Figure 5). As an alternative to IPW, ARC/INFO AML routines (Rich *et al.*, 1994) can be used to calculate topographic shading of direct beam shortwave radiation. During the hydrological model simulation, the map of calculated incoming direct beam radiation is adjusted to produce actual surface short-wave radiation based on measured or inferred cloud cover. Diffuse surface radiation is assumed to be unaffected by cloud cover.

Precipitation maps can be constructed using either radar images, interpolated gauge data or model simulations [such as the orographic precipitation model of Rhea (1978)] to define the spatial distribution of precipitation over the basin. At a minimum, the meteorology required by DHSVM is limited to incoming short- and long-wave radiation, air temperature, relative humidity, wind speed and precipitation at one or more points in or near the watershed. Meteorology is interpolated between observation stations. Pre-processing of the base meteorological data is limited to estimation of missing measurements. If there is only one observation station in or near the basin, DHSVM automatically adjusts temperature and humidity for elevation via pseudo-adiabatic lapse rates.

Land surface characterization

DHSVM characterizes the land surface at the spatial scale of the DEM. Land surface information is needed to characterize the physical characteristics of the multiple soil layers as well as vegetation understorey and overstorey properties. With respect to soil properties, DHSVM requires porosity, field capacity, wilting point and vertical hydraulic conductivity for each pixel. However, instead of specifying a soil map for every soil parameter, DHSVM categorizes each pixel based on the dominant soil type, classification or parent material. The specific soil attributes for each pixel are then obtained from a look-up table referenced by soil index. For example, a pixel with predominately alluvial soils can be given different parameters to a soil with a volcanic origin; however, all soils of alluvial origin will have the same parameters, unless subclasses are specified. In our applications, soil data were obtained from the USFS Soils Inventory (USFS, 1975), the STATSGO soils database (USDA, 1994) or were provided as GIS layers from private landowners.

The distribution of vegetation over the watershed is described in a similar manner. DHSVM requires information on a large number of attributes for both the overstorey and understorey such as leaf area index, height, stomatal conductances and radiation/wind speed decay coefficients. Many of these factors depend not only on the vegetation type, but also on its age. Therefore, for each pixel, the dominant vegetation type for each layer is defined.

Vegetation type is used to access the appropriate vegetation parameters for each storey in a look-up table. Adjustments to the parameters are made based on the age of the vegetation as calculated from the current simulation year minus the year when the vegetation was planted or cut (its year of origin). For our application, dominant overstorey vegetation type and its year of origin were obtained from GIS-based vector maps and attribute tables provided by public land owners such as the USFS and Washington State Department of Natural Resources. As an alternative, classified remote sensing imagery can provide the necessary vegetation information. Examples of dominant overstorey species and year of origin are shown in Figures 2 and 6.

Road networks are described in a manner similar to the stream channel network. If previous surveys of the road network in the basin are available, these data can be used directly to locate the road network on the DEM. However, if unavailable, these data must be collected. The availability of inexpensive, portable geographic positioning system (GPS) receivers makes definition of the road network possible by simply traversing the network. Once the road network is located, DHSVM requires information on road attributes that affect the fate of intercepted water on each segment. These attributes include: representative road surface width, roughness, cut bank slope, cut bank height, ditch depth and ditch width. Each road segment is further categorized as in sloped, out sloped or crowned. Out sloped segments discharge intercepted water directly to the hillside. In sloped segments discharge water to the culvert. Information on the 'from' and 'to' nodes are also required for the road network.

Initial model state

DHSVM also requires initial conditions for each of its hydrological state variables. These variables include the distribution of snow and water on both vegetation layers, the ground surface and in all soil layers. Uniform distributions of these parameters are typically selected for the initial simulation. If the initial conditions are not well known, the model can be run for a 'warm-up' period, which is subsequently discarded. Furthermore, at any point in a simulation, DHSVM can output its hydrological state variables. These saved states can then be used to restart the model.

Output

Given the land surface characteristics data and meteorological time series, DHSVM can simulate any watershed independent of the GIS platform used to construct the input files. During the simulation, DHSVM tracks hydrological variables and fluxes such as snow water equivalent, depth to saturation, soil moisture in two subsurface zones and moisture intercepted on the vegetation as snow or liquid. Depending on the application, DHSVM can be tailored to only output maps of the variables of interest. These maps are then easily imported into GIS packages for synthesis and analysis.

Calibration

Once the input files are complete, DHSVM is calibrated to a multi-year period of observed stream flow and snow cover record. Typically, DHSVM is 'spun-up' for at least a one-year period (including a full snow season with summer dry down). This period is not used in the calibration and is necessary so that the initially specified distribution of soil moisture over the watershed does not bias results during the calibration and/or simulation period. The average annual water balance is used to adjust the observed precipitation record. Calibration to match observed basin outlet hydrographs has generally been limited to adjustment of selected parameters that are assumed constant over the catchment (or for each soil type), such as the depth of the rooting soil layers, depth of soil below the rooting zones and lateral saturated hydraulic conductivity. Calibration of the distributed velocity or channel routing algorithms is accomplished by matching observed and predicted hydrographs for select peak stream flow events. If available, observed data on the distribution of soil moisture and snow cover should be used to calibrate DHSVM further. In many applications, this additional calibration would be limited to data available from NRCS SNOTEL sites. Unfortunately, SNOTEL data are collected in open areas, which precludes local calibration of those vegetation parameters affecting snow accumulation and melt for individual watersheds. These parameters should be selected with care based on local climate effects on vegetation and hydrological limitation of the parameters' importance. For example, the canopy attenuation coefficient for short-wave radiation is not critical for prediction of peak stream flows in Western Cascade watersheds since these events are not radiation dominated. However, in east-side basins, this parameter is of critical importance in predicting peak events. For our applications, DHSVM has been extensively verified against observed snow accumulation and ablation records collected under mature Douglas-fir stands in the central Oregon Cascades (Storck et al., 1997). In our experience, calibration of DHSVM to a new basin after application to a basin of similar geomorphology is limited to those parameters, such as canopy coefficients, that are expected to vary between basins of different climate.

APPLICATIONS

To illustrate the use of DHSVM to investigate the effect of land use changes related to forest harvesting on watershed hydrology, three Cascade Mountain catchments were studied. All three of these basins (see Figure 2) have been heavily harvested over the last 50 years. Figure 2 shows the location, elevation, dominant overstorey vegetation species and the year of origin for each pixel in the application watersheds. (Note that for Hard and Ware creeks, which are owned entirely by Weyerhaeuser, specific vegetation species and year of origin are not shown. Instead, the raw vegetation classes used in the attribute look-up table are shown to illustrate spatial variability of species and age.) These three sites were selected because each provides a unique set of climatic influences and harvest (or road construction) histories.

Two of these basins, the North Fork of the Snoqualmie (165 km^2) and Hard/Ware Creek (5·2 km^2) drain the western slopes of the Cascades. These basins are headwaters of the Snoqualmie and Deschutes rivers, respectively. The elevation of the North Fork Snoqualmie catchment ranges from 350 to 1800 metres while the elevation of Hard and Ware Creeks ranges from 460 to 1200 metres. This elevation range places both basins in the west Cascades transient snow zone and therefore their peak stream flow hydrology is dominated by winter rain-on-snow events. The Little Naches River catchment drains 388 km^2 on the east slope of the Washington Cascades with an elevation range from 800 to 1900 metres. Its hydrology is dominated by spring snowmelt events, although it does experience occasional rain-on-snow events as well. Application of DHSVM on the North Fork Snoqualmie focused on prediction of rain-on-snow peak stream flows and enhancement owing to forest harvesting. The climatology of Hard and Ware Creeks is similar to that of the North Fork Snoqualmie; the study of the former focused on the effects of forest roads on runoff peaks. Application of DHSVM to the Little Naches River focused on prediction of peak spring stream flow events and enhancement owing to forest canopy removal.

Landowners in the Little Naches and North Fork Snoqualmie catchments include the US Forest Service, Plum Creek Timber Company and the Weyerhaeuser Company. The sole landowner of Hard and Ware Creeks is the Weyerhaeuser Company. These landowners were the primary source of data describing the distribution of vegetation over the watersheds shown in Figure 2.

North Fork Snoqualmie River watershed

DHSVM was applied to the North Fork Snoqualmie catchment as part of a retrospective analysis of peak stream flow events from 1948 to 1993. Results presented here are for a select analysis of stream flow events from October 1989 to December 1990. This period includes the second largest flood of record on the North Fork Snoqualmie which occurred during a rain-on-snow (ROS) event in November 1990 (observed maximum discharge 330 m^3 s^{-1}) and another major ROS event in December 1989 (observed maximum discharge 226 m^3 s^{-1}). Because of the relatively short period of analysis, the results may not be representative in a statistical sense. However, they do serve to illustrate possible increases in stream flow and meltwater production for two actual events.

The results presented are for the hydrograph difference series of two simulations with different vegetation scenarios. The first simulation used the current (1996) vegetation cover. The second assumed that the entire basin was clear-cut. Although the latter simulation is unrealistic, it does serve to set the upper bound of expected additional forest harvest contributions to increases in stream flow. Furthermore, information on the spatial distribution of snowmelt increases obtained from such a simulation serves to identify areas in the watershed that are most likely to yield increased snowmelt as a result of forest harvesting for particular storm events. To ensure accurate representation of hydrograph peaks, all simulations were conducted using a 3-hour time step.

The hydrograph difference series of these two simulations (clear-cut minus current vegetation stream flow) is shown in Figure 3 to illustrate the expected seasonal effect of forest harvesting on stream flow. The raw difference in each 3 hour predicted stream flow is shown ordered by date. Maximum increases in stream flow are predicted to occur during ROS events. The ROS event of 4 December 1989 would increase by 70 m^3 s^{-1}

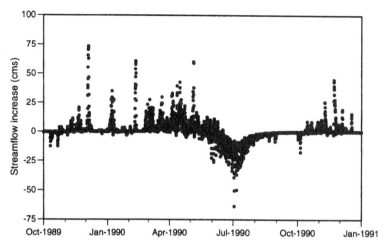

Figure 3. Predicted increase in stream flow due to complete forest harvesting of the North Fork Snoqualmie basin. Flood of record is 300 m^3 s^{-1}

(approximately 31%) owing to complete clear-cutting, while the larger ROS event of 24 November 1990 would increase approximately 30 m^3 s^{-1} (or 10%). Figure 3 also shows similarly large predicted increases in stream flow during the early part of the spring snowmelt season in April and the first half of May. This increase is followed by a pronounced decrease in stream flow during the snowmelt period from June through to July.

The mechanisms for the changes in peak stream flow and the differences between the predicted increase for both ROS events can be understood by examining the spatial distribution of snow water equivalent and snow melt during these events. Figure 4 shows the predicted distribution of snow water equivalent before and after the ROS event of 4 December 1989 for both the entirely clear-cut and current vegetation simulations. Also shown are the differences in snowmelt (clear-cut minus vegetated) during the height of the runoff production as well as the DEM. During the December ROS event, the average snow water equivalent (SWE) over the basin decreased from 0·25 m on 3 December to 0·20 m by 4 December for the entirely clear-cut simulation. For the current vegetation scenario, the predicted average SWE decreased from 0·28 to 0·27 m. The greater initial SWE for the current vegetation scenario is due to sheltering of low elevation snowpacks from a previous smaller melt event. For the clear-cut simulation, the decrease in SWE over the basin was mostly due to rapid ablation of low elevation snow packs. These low elevation snow packs were sheltered during the vegetated simulation and thus a large increase in peak stream flow was predicted for the clear-cut simulation.

Although not shown on the figure, snowpack dynamics were considerably different during the ROS event of November 1990. The average basin snow water equivalent increased from 0·08 to 0·10 m from 24 November to 25 November for the entirely clear-cut simulation, and from 0·10 to 0·12 m for the vegetated simulation. The November event was also characterized by little or no snow at low elevation prior to the event. Thus, the predicted increase in peak stream flow during the November 1990 event was much smaller than during the December 1989 event, because increases in snowmelt were limited to a much narrower elevation band over the basin. These results show the importance of antecedent conditions on predictions of forest harvest effects. In the context of watershed analysis, these results identify the lower elevation portions of the basin as contributing the most towards increases in snowmelt owing to forest harvesting. In fact, the snow accumulation images show increases at the highest elevations as a result of the absence of forest interception.

Figure 5 shows the distribution of snow melt during a typical radiation dominated melt event on 4 April 1990. Snowmelt is shown for both the entirely clear-cut and current vegetation simulations. Also shown is the difference image of snowmelt over the basin. For the entirely clear-cut case, the figure shows that meltwater production is largely determined by aspect. During the morning hours, maximum snowmelt occurs on the east-facing slopes. Maximum production is seen on south-facing slopes during midday and shifts to west-facing slopes by late afternoon. However, for the current vegetation simulation, melt is controlled as much by vegetation distribution as it is by aspect. The difference image clearly shows that high elevation, south-facing slopes are the major contributor to increases in peak stream flows during early season, radiation-dominated snowmelt events. These are precisely the slopes that are relatively unaffected by ROS events.

The pronounced decrease in stream flow due to forest harvesting during the later summer months seen in Figure 3 is also primarily a result of changes in snow distribution. After clear cutting, much more snowmelt occurs during the early part of the spring season than in the vegetated simulation. This early increase in melt leads to a rapid reduction in snowpack and, subsequently, less snow available for late season melt.

Little Naches River

DHSVM was applied to the Little Naches River watershed to evaluate its capability to perform a hydrology assessment of the type required by the Washington State Forest Practice Board (WFPB, 1995). The intent of WFPB hydrology assessments is to identify hydrologically sensitive areas (which may include the entire watershed or subareas within it) in watersheds where logging is proposed. Land surface data for the Little Naches catchment were supplied by the major landowners in the basin, which include Plum Creek Timber Co. and the US Forest Service (USFS). Vegetation data were provided by Plum Creek Timber Co. Soil data were obtained from the USFS Soils Inventory (USFS, 1975).

Hydrological sensitivity to timber harvesting was evaluated by performing model runs for selected water years and peak flow events with prescribed forest canopy scenarios. Model runs were made for the current vegetation condition, a mature even-aged forest and two possible harvest scenarios: a 'maximum potential' (which corresponds to clear cutting all Plum Creek Timber Co. lands, and all but 15% of the USFS lands); and a 15% dispersed retention, which corresponds to current vegetation on all US Forest Service lands, and logging of all but 15% of the Plum Creek Timber Co. lands. The overstorey age for the harvest prescriptions and the current vegetation condition are shown in Figure 6 along with the catchment topography. The current vegetation distribution is based on 1996 data.

Since this application of DHSVM focused on predicting increases in spring melt events, the model was calibrated to the observed stream flow of water year 1993. During this year, the hydrograph was dominated by a single spring snowmelt peak. Calibration parameters obtained from the previous application to the North Fork Snoqualmie catchment were used to the extent possible; additional calibration focused on selection of canopy attenuation coefficients to reflect timing of the observed spring hydrograph peak for 1993.

The results of our analysis are presented in Table I, which shows the predicted increase for two spring snowmelt events (May 1993 and April 1994) over the entire basin and for a smaller (approximately 20 km² drainage area) headwaters catchment within the Little Naches, Pyramid Creek. Percentage increases are calculated relative to the simulations for 100% mature forest canopy. The simulation showed that current levels of forest harvesting should lead to only minor increases in peak spring stream flow of approximately

Table I. Summary of similated peak flow changes by vegetation scenario

Basin	Percentage increase in peak flow			
	Event date	Current condition	Maximum potential harvest	15% dispersed retention
Naches WAU	May 1993	3·3	30·5	7·1
Naches WAU	April 1994	2·8	20·8	5·8
Pyramid Creek	May 1993	10·7	30·4	14·5
Pyramid Creek	April 1994	17·8	30·0	21·2

3% over the entire basin. However, increases of near 30% would occur for the maximum possible harvest scenario. Increases less than 10% would occur for the more realistic 15% dispersion harvest. For the Pyramid Creek catchment, larger increases in peak stream flow are predicted. For example, current harvest levels should lead to increases in stream flow as high as 18% relative to mature vegetation. A 15% dispersed retention harvest on Plum Creek Timber Co. land in Pyramid Creek would increase peak stream flow by approximately 20% relative to a fully mature forest, or only 2% points above the current condition.

The spatial distribution of increases in snowmelt with respect to vegetation scenarios is used to determine the hydrological sensitivities to harvest. Figure 7 shows the distribution of increases in snowmelt based on the maximum potential harvest scenario relative to the mature vegetation simulation. Differences are shown for snowmelt production in the seven days prior to the peak spring events of May 1993 and April 1994. Areas with no snowmelt in either simulation are not coloured. In the May 1993 event, snowmelt was limited to higher elevations and the total area contributing increases in snowmelt was significantly larger than the area contributing decreases owing to harvesting. Areas providing less snowmelt are limited to lower elevations. The figures show that the high elevation headwaters of each sub-basin are most likely to contribute increased melt during spring peak flows. For April 1994, almost the entire basin contributed snowmelt to the peak flood event. However, the areas of increase in the headwaters were countered by decreases in expected snowmelt at lower elevation. Although more of the basin contributed melt in the April 1994 event, predicted increases resulting from forest harvesting would be greater during the May 1993 event. Exactly the opposite trend is seen in Pyramid Creek. For the April 1994 event, much more of Pyramid Creek contributes to increased snowmelt than during the May 1993 event. Consequently, predicted increases for Pyramid Creek are greater for the April 1994 event. In the context of a watershed analysis, these results suggest that, for the antecedent snow conditions and event characteristics simulated, the high elevation headwaters of the Little Naches catchment are much more sensitive to timber harvest than are the lower elevations.

Hard and Ware Creek catchments

To evaluate the effect of forest roads on peak flows, DHSVM was applied to Hard and Ware Creeks, two headwater catchments of the Deschutes River in western Washington. Stream channels were determined by specifying a minimum contributing area of 2 hectares to approximate field observations of stream network extent during high flow winter runoff events. The derived stream network was compared with field observations of channel extent and culvert location and modified accordingly. The depth and width of the channel incision into the soil layer was specified based on channel order and visual observations.

Ditch-relief and stream-crossing culverts in the Hard and Ware creeks watersheds were located using a hand-held portable global positioning system (GPS). Position data were differentially corrected for an estimated horizontal precision of 2–5 m. A total of 109 culverts were located: 65 in the Hard Creek basin and 44 in the Ware Creek basin. Current road locations were surveyed on foot using GPS with a sampling frequency of 5 seconds. A total of 22·1 km of roads were located in the two basins, 10·7 km in Ware Creek and 11·4 km in Hard Creek.

Culverts were classified in the field according to their connectivity to the natural drainage system. The extended stream network due to roads includes road segments draining to connected culverts and gullies. The connection of culverts to natural stream channels was determined by following the path of each gully, during average winter runoff conditions, until a stream channel was encountered or infiltration occurred. The locations of the culverts, road segments and the extended stream channel network are shown on Figure 8. Natural stream channels are shown in grey with extensions resulting from the road network being shown in black. The length of road providing direct runoff into the stream system amounts to 52% of the natural channel length in Ware Creek, and 63% in Hard Creek. These increases are similar to those of Wemple *et al.* (1996) and Montgomery (1994) who reported increases of 40% and 23–57%, respectively. Our methods were similar to both previous investigations, except that Wemple *et al.* (1996) assumed that culverts were connected through gullies if a gully extended for at least 10 m below a culvert, and Montgomery (1994) did not take into account the length of any new channels incised between culverts and channel heads.

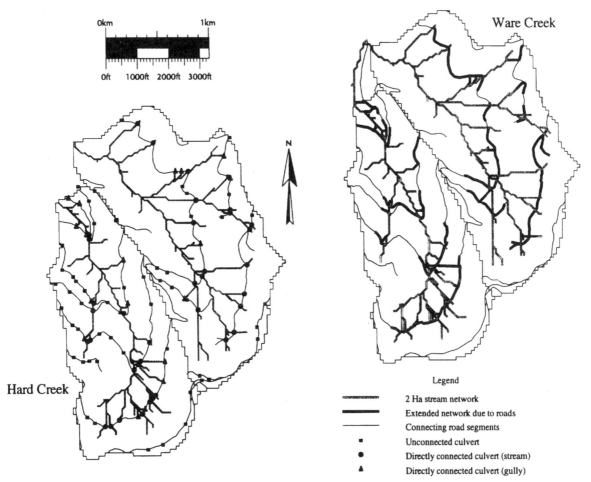

Figure 8. Culvert locations and extended stream network due to roads, Hard and Ware creeks, Washington

DHSVM was calibrated with the imposed road and stream network algorithms to observed stream flow for water year 1991. The hydrograph for this year includes several ROS peaks. Calibration focused on adjustment of soil parameters to control basin responsiveness to individual rainfall events.

To isolate the effect of roads, the percentage increase in simulated stream flow with the imposed road network was calculated relative to simulated stream flow without the road network for water year 1992. Simulated hydrographs for a typical storm occurring in November 1991 are shown in Figure 9. These hydrographs indicate a slight decrease in time to peak, an increase in peak discharge and an increased rate of recession because of the road network. This pattern is typical of other simulated storms for isolated peaks of the first peak in a series. The average increase in peak flow for the five largest events in 1992 was 17·4% for Hard Creek and 16·2% for Ware Creek. The maximum percentage increase of 27% occurred in Hard Creek for the fifth largest storm of water year 1992 (20 November 1991). Although not directly comparable, these results are similar to those of Jones and Grant (1996) who found an increase in the mean peak discharge owing to roads alone of 14%.

In general, the results of this analysis indicate an increase in peak stream flow as a consequence forest roads. The average observed increase was larger for Hard Creek, the basin with the greater percentage increase in drainage density owing to roads.

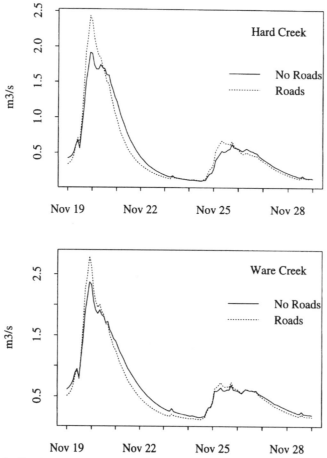

Figure 9. Predicted Hard and Ware creek discharge with and without road network, November 1991

CONCLUSIONS

Distributed hydrological modelling can be a powerful tool for investigation of the hydrological effects of land use change. The three applications described here illustrate how such a model can be linked with modern GIS platforms to predict the effects of forest harvesting on watershed hydrology. DHSVM requires a detailed description of the distribution of physical parameters affecting the water and energy balance at the land surface. Because of these data input requirements, the method is facilitated by use and processing of GIS vegetation, topographic and soil data layers. Model input files are derived directly from GIS layers, and model output is readily processed and interpreted using GIS software.

The three examples, for Cascade Mountain watersheds that have been moderately to heavily harvested, illustrated how the distributed model output maps can provide insights into hydrological changes in a watershed, and, more importantly, how areas within the watershed that contribute most to these changes can be identified. For instance, the North Fork Snoqualmie catchment, which is dominated by rain-on-snow peaks, is more sensitive to forest harvest at low to mid elevations than is the Little Naches watershed, where stream flow peaks occur during spring snowmelt, and for which high elevations are most sensitive. The applications also illustrate the importance of antecedent conditions for prediction of forest harvesting effects. Changes in peak stream flow are particularly sensitive to the initial distribution of snow cover over a basin. The Hard and Ware creeks examples show the importance of hydrograph timing for the prediction of road network effects: isolated hydrographs are more sensitive to peak flow increases than are secondary hydrographs following an extended storm period. However, roads lead to faster basin response time in both instances.

ACKNOWLEDGMENTS

The research on which this paper was based was supported by an Earth Science Fellowship from the National Aeronautics and Space Administration (to P.S.), and by contracts from the State of Washington Department of Natural Resources (FY96-088), and the National Council of the Paper Industry for Air and Stream Improvement (CWE-14), to the University of Washington. The authors are grateful to the Weyerhaeuser Co. for providing vegetation, soils and hydrometeorological data, as well as access for field work, to Hard and Ware creeks, and to the Plum Creek Timber Co. for providing vegetation and soils data for the Little Naches River.

REFERENCES

Anderson, E. A., 1968 'Development and testing of snowpack energy balance equations', *Wat. Resour. Res.*, **4**, 19–37.

Anderson, H. W. and R. L. Hobba, 1959. 'Forests and floods in the northwestern United States', *IAHS Publ.*, **48**, 30–39.

Berris, S. N. and R. D. Harr, 1987. 'Comparative snow accumulation and melt during rainfall in forested and clear-cut plots in the western Cascades of Oregon', *Wat. Resour. Res.*, **23**, 135–142.

Bowling, L. C. and D. P. Lettenmaier, 1997. 'Evaluation of the effects of forest roads on stream flow in Hard and Ware Creeks, Washington, Water Resources Series', *Technical Report No. 155*. University of Washington, Seattle.

Bunnel, F. L., McNay, R. S., and Shank, C. C. 1985. 'Trees and snow: the deposition of snow on the ground — a review and quantitative synthesis', *IWIFR-17*. Research Branch, Ministries of Environment and Forests, Victoria, BC.

Burnash, R. J. C., Ferral, R. L., and McGuire, R. A. 1973. 'A generalized stream flow simulation system — conceptual modeling for digital computers', US National Weather Service, Joint Federal-State Forecast Center, Sacramento.

Calder, I. R. 1990. '*Evaporation in the Uplands*. John Wiley & Sons, Chichester. 148 pp.

Christner, J. and Harr, R. D. 1982. 'Peak stream flows from the transient snow zone, western Cascades, Oregon', in Bernard Shafer (ed.), *Proceedings of the 50th Western Snow Conference, Colorado State Univ. Meeting, Reno, Nevada, Fort Collins, CO*. pp. 27–38.

Connelly, B. A., Cundy, T. W., and Lettenmaier, D. P. 1993. 'Implications of forest practices on downstream flooding: Phase I Interim Report', Washington Forest Protection Association, Olympia, WA, unpublished.

Crawford, N. H. and Linsley, R. K. 1966. 'Digital simulation in hydrology: Stanford watershed model IV', *Technical Report No. 39*. Department of Civil Engineering, Stanford University.

Dubayah, R., Dozier, J., and Davis, F. W. 1990. 'Topographic distribution of clear-sky radiation over the Konza Prairie, Kansas', *Wat. Resour. Res.*, **26**, 679–690.

Entekhabi, D. and Eagleson, P. S. 1989. 'Land surface hydrology parameterization for atmospheric general circulation models: inclusion of subgrid scale variability and screening with a simple climate model', *Rep. 325, Ralph M. Parsons Lab*. Mass. Inst. of Technology, Cambridge. 195 pp.

Frew, J. E. 1990. 'The image processing workbench' *PhD Thesis*. Department of Geography, University of California, Santa Barbara. 382 pp.

Harr, R. D. 1981. 'Some characteristics and consequences of snowmelt during rainfall in western Oregon', *J. Hydrol.*, **53**, 277–304.

Harr, R. D. 1986. 'Effects of clear-cut logging on rain-on-snow runoff in western Oregon: a new look at old studies', *Wat. Resour. Res.*, **22**, 1095–1100.

Jones, J. A. and Grant, G. E. 1996. 'Peak flow responses to clear-cutting and roads in small and large basins, western Cascades, Oregon', *Wat. Resour. Res.*, **32**, 959–974.

Longley, K., Jacobsen, D., and Marks, D. 1992. 'Supplement to the Image Processing Workbench (IPW) report', *Environmental Research Lab*. US Environmental Protection Agency, Corvallis, Oregon.

Maidment, D. R., Olivera, J. F., Calver, A., Eatherall. E., and Fraczeck, W. 1996. 'A unit hydrograph derived from a spatially distributed velocity field', *Hydrol. Process.*, **10**, 831–844.

Montgomery, D. R. 1994. 'Road surface drainage, channel initiation and slope stability', *Wat. Resour. Res.*, **30**, 1925–1932.

Montgomery, D. R., Grant, G. E., and Sullivan, K. 1995. 'Watershed analysis as a framework for implementing ecosystem management', *Wat. Resour. Bull.*, **31**, 1–18.

Moore, I. D., Grayson, R. B., and Ladson, A. R. 1991. 'Digital terrain modelling: a review of hydrological, geomorphological, and biological applications', *Hydrol. Process.*, **5**, 3–30.

Perkins, W. A., Wigmosta, M. S., and Nijssen, B. 1996. 'Development and testing of road and stream drainage network simulations within a distributed hydrologic model', *Eos Trans. AGU*, **77**(46) (Fall Meet. Suppl.), F453.

Rhea, J. O. 1978. 'Orographic precipitation model for hydrometeorological use', *Atmospheric Science Paper No. 287*, Colorado State University, Fort Collins, CO.

Rich, P. M., Hetrick, W. A., and Saving, S. C. 1994. 'Modeling topographic influences on solar radiation: a manual for the SOLARFLUX model (Draft)'. University of Kansas, Lawrence.

Schmidt, R. A. and Troendle, C. A. 1992. 'Sublimation of intercepted snow as a global source of water vapor', in Charles Troendle (ed.), *Proceedings, 60th Western Snow Conference, Jackson, Wyoming. Colorado State University, Fort Collins, CO. pp 1–9.*

Storck, P., Lettenmaier, D. P., Connelly, B. A., and Cundy, T. W. 1995. 'Implications of forest practices on downstream flooding: Phase II Final Report', *TFW-SH20-96-001.* Washington Forest Protection Association, Olympia, WA. 100 pp.

Storck, P., Kern, T., and Bolton, S. M. 1997. 'Measurement of differences in snow accumulation, melt, and micrometeorology between clearcut and mature forest stands', in Charles Troendle (ed.), *Proceedings, 65th Western Snow Conference, Bonff, Alberta, Colorado State University, Fort Collins, CO.*

USDA (US Department of Agriculture) 1994. *State Soil Geographic (STATSGO) Data Base, Miscellaneous Publication 1492.* Soil Conservation Service, United States Department of Agriculture, Washington DC.

USFS (US Forest Service) 1975. *Soil Resource Inventory.* Yakima Ranger District, Yakima, WA.

Wemple, B.C., Jones, J. A., and Grant, G. E. 1996. 'Channel network extension by logging roads in two basins, Western Cascades, Oregon', *Wat. Resour. Bull.,* **32**, 1195–1207.

WFPB (Washington State Forest Practices Board) 1995. 'Standard methodology for conducting watershed analysis', *Washington Forest Practice Act Board Manual, Version 3.0.* Olympia, WA.

Wigmosta, M. S., Lettenmaier, D. P., and Vail, L. W. 1994. 'A distributed hydrology-vegetation model for complex terrain', *Wat. Resourc. Res.,* **30**, 1665–1679.

7

MODELLING RUNOFF AND SEDIMENT TRANSPORT IN CATCHMENTS USING GIS

A. P. J. DE ROO*

Department of Physical Geography, Utrecht University, PO Box 80.115, 3508TC Utrecht, The Netherlands

ABSTRACT

The application of geographical information systems (GIS) in modelling runoff and erosion in catchments offers considerable potential. Several examples illustrate simple GIS techniques to produce erosion hazard indices or erosion estimates using USLE-type models. Existing erosion models can also be loosely coupled to a GIS, such as the ANSWERS model. Furthermore, models can be fully integrated into a GIS by embedded coupling, such as the LISEM model. However, GIS raster-based erosion models do not necessarily produce better results than much simpler and partly lumped erosion models with 'representative elements', although they reproduce topography in more detail. The reasons for the disappointing results of spatial models must be sought in the uncertainty involved in estimating and measuring the large number of input variables at a catchment scale. There is a need for much simpler loosely coupled or embedded GIS erosion models simulating only the dominant processes operating in the catchment.

KEY WORDS   soil erosion models; runoff catchment; GIS; LISEM; PCRaster

INTRODUCTION

In many areas of the world flooding and soil erosion are major problems. In the intensively used agricultural area of the loess belt in north-western Europe, both flooding and erosion cause damage to private property (Boardman *et al.*, 1994). In many catchments in mountainous areas, such as the Alps and the Pyrenees, flash-floods and the associated soil erosion are a major problem, e.g. in the Ouveze catchment in France (Amarú *et al.*, 1996; Wainwright, 1996). The prediction of peak flows, total flows, source areas and soil erosion and deposition amounts is necessary for understanding the problem, designing control measures and developing and evaluating scenarios that can reduce both flooding and erosion. Flooding and erosion are influenced by many factors such as rainfall distribution, soil factors and land use-related factors. Since all these factors vary in both space and time, the use of geographical information systems (GIS) offers considerable potential. The aim of this paper is to present both simple and complex methods for predicting soil erosion risk in catchments using GIS techniques. Various GIS approaches are discussed and illustrated by applying them to the Catsop catchment (South-Limburg, The Netherlands).

In the loess area of South-Limburg, The Netherlands, three research catchments have been continuously monitored since 1987. One of these catchments is the Catsop catchment, situated between the villages or Elsloo and Beek (De Roo, 1993, 1996). The catchment covers 42 hectares and also has a gently to moderately sloping topography with altitudes ranging from 80 to 110 m above sea level. About 86% of the slopes have a

---

*Correspondence to: A. P. J. De Roo, Joint Research Centre, Space Applications Institute, Agriculture Information Systems Unit, Ispra, TP 441, I-21020 Ispra (VA), Italy.

gradient of less than 10%, and 3·5% of the slopes are steeper than 15%. The dominant flow direction of surface runoff is to the west. The parent material is loess. The loess deposits lie on top of Tertiary sand deposits and Quaternary deposits of the 'West-Maas' River. Both deposits locally appear close to the soil surface. All soils are developed in loess.

The land use in the Catsop catchment is dominated by arable cultivation with three important crops: winter wheat, sugar beet and potatoes. During recent years, maize has also been grown in one field. Thirteen per cent of the catchment is under permanent pasture. Mean annual precipitation is estimated at 675 mm. Precipitation, discharge and sediment loadings are measured automatically at the catchment outlet. During major storms in the Catsop catchment 3–30% of total rainfall reaches the catchment outlet (De Roo, 1993).

USING GIS IN SOIL EROSION RISK ASSESSMENT

GIS can be used to provide a rapid assessment of erosion hazards and amounts. Many authors have demonstrated the potential for using digital elevation models (DEM) in soil erosion assessment (Burrough, 1986; De Roo et al., 1989; Moore et al., 1992, 1993). Since, at present, the most detailed information available for catchment modelling is information about the terrain in the form of distributed elevation data, a catchment should be modelled using the DEM. Using GIS techniques several products can be calculated from a DEM that are very useful for soil erosion modelling. The most important of these products are:

(i) slope gradient, a dominant factor in the erosion process, controlling the overland flow rate;

(ii) aspect or local drainage direction, controlling the direction of the flow down slope; and

(iii) contributing or upstream area of a point in the landscape, controlling potential discharge.

In the mid-1980s, GIS were used to calculate soil erosion over larger areas using the universal soil loss equation (USLE) or slight modifications of the USLE. At Utrecht University in the Netherlands a GIS version of the USLE (Jetten et al., 1988) was used in the Ardèche region in France (Van Hees et al., 1987) and in South-Limburg in the Netherlands (De Roo et al., 1987). The USLE can be expressed as

$$A = R \cdot K \cdot LS \cdot C \cdot P \tag{1}$$

where A = soil loss (tons/ha/year), R = rainfall erosivity factor, K = soil erodibility factor, LS = slope length and slope steepness factor, C = crop factor, and P = management factor.

An example of its application to the Catsop catchment in South-Limburg (The Netherlands) is provided below. From the DEM (Figure 1), a slope gradient map (Figure 2) can be calculated. The calculation of the slope length factor is a little more complicated. From the point of interest one searches towards the watershed boundary or to the nearest 'flat' area above this point. Calculation of the USLE in a GIS is easy, since it involves multiplication of six maps, and multiplication is a standards option in most GIS. Only the calculation of slope length is not a standard option. Bork and Hensel (1988) also used a GIS USLE approach in Germany to estimate soil erosion amounts over larger areas. However, the USLE has not been developed for use at the catchment scale, since it does not simulate deposition. Ferri and Minacapilli (1995) extended the USLE with a sediment delivery term such that the model could predict sediment output from the catchment. As an example of annual soil loss in the Catsop catchment, estimated using the original USLE, the actual land use of 1993, with a K-factor of 0·50, an R-factor of 70 and a P-factor of 1, is given in Figure 3. Comparing Figures 2 and 3 it can be seen that slope gradient is a dominant factor in this catchment.

At the catchment scale, however, the slope length approach is difficult to use. Above every point, 'contributing' or 'upstream' area rather than slope length is the key determining factor (Moore et al., 1993; Desmet and Govers 1996). Since this is also easy to calculate within a GIS, a modified USLE can be evaluated. A procedure for estimating the LS-factor using contributing area is provided by Moore et al. (1993)

$$LS = (n + 1)\left[\frac{A_s}{22\cdot13}\right] \cdot^n \left[\frac{\sin\beta}{0\cdot0896}\right]^m \tag{2}$$

(a)

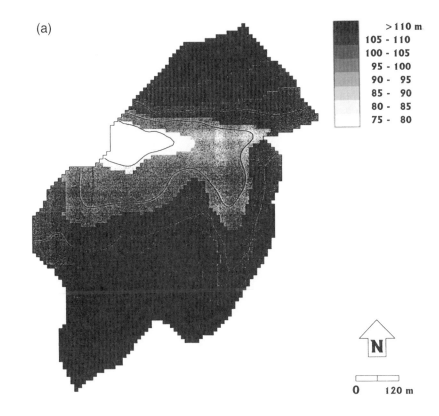

(b)

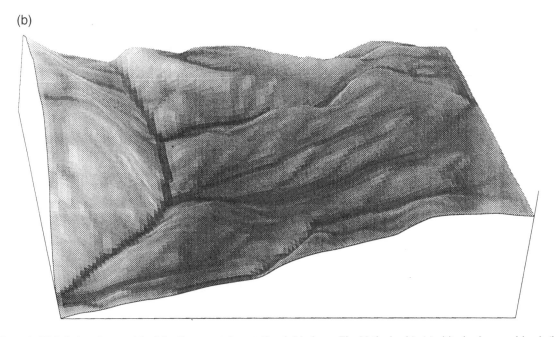

Figure 1. Digital elevation model of the Catsop catchment (South-Limburg, The Netherlands): (a) altitude above seal level, (b) the wetness index (see text)

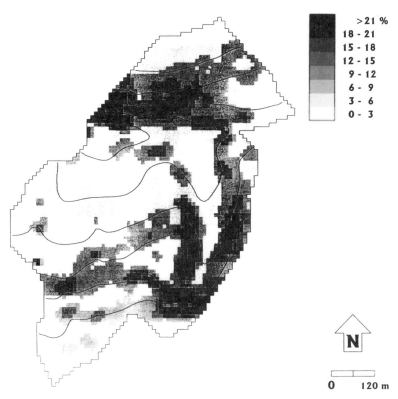

Figure 2. Slope gradient in the Catsop catchment, calculated using the PCRaster GIS

where LS = slope length factor, n = constant (0·4), β = local slope gradient (degrees), m = constant (1·3) and A_s = unit contributing area.

Contributing area can also be easily calculated within the PCRaster GIS. Using the LS formula, a map can be produced (Figure 4). This LS map can be used as a potential erosion hazard map. Again it is clear that slope gradient is a dominant factor. Using this LS-factor, the improved USLE can be applied at a catchment scale (Figure 5). However it must be recognized that the USLE does not predict deposition. Especially in valley bottoms, where occasionally high LS values are calculated (e.g. Figure 4, near the catchment outlet), significant deposition can occur. Also, because the USLE is mainly an interrill erosion prediction model, concentrated rill and gully erosion are not predicted, or at least are underpredicted, leading to underprediction of soil loss amounts at the catchment scale. Other potential applications of GIS-based USLE soil erosion modelling for large areas are provided by Fröhlich et al. (1994) and Dräyer (1996).

Moore et al. (1992, 1993) provided some examples of other indices useful for soil erosion hazard mapping, calculated using the basic products of digital elevation models such as slope gradient (S) and contributing area (A_s). Examples are the wetness index, a stream power index and a sediment-transporting capacity index. The wetness index (Beven and Kirkby 1979) gives an idea of the spatial pattern of soil moisture content

$$Wetness = \ln\frac{A_s}{S} \tag{3}$$

where A_s = unit contributing area (m^2/m) and S = slope gradient (m/m)(tan $\beta = S$).

Figure 6 shows the spatial distribution of the wetness index in the Catsop catchment. Logically, the valley bottoms have high wetness values. Thus, the wetness index can be used to identify possible stream paths. Its most important use, however, is the indication of 'wet' and 'dry' areas, and thus possible source areas for saturation overland flow, which produces erosion. The wetness index could also be used as a covariable

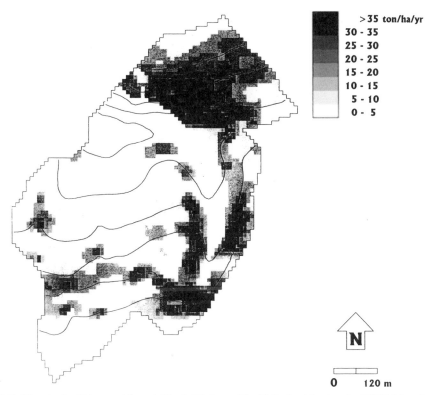

Figure 3. Soil loss in the Catsop catchment (South-Limburg, The Netherlands) using the USLE (slope length version)

together with point field measurements of soil moisture, to provide a basis for spatial interpolation using, for example, co-Kriging.

The stream power index (Moore *et al.*, 1992) is the first of several indices indicating possible erosion by concentrated flow

$$\Omega = A_s \cdot S \tag{4}$$

where Ω = stream power, S_s = unit contributing area (m^2/m) and S = slope gradient (m/m)(tan $\beta = S$). Combined with soil and land use factors (e.g. the K and C-factors), the stream power index gives a good indication for flow detachment risk (Figure 7). In a similar way, a somewhat modified version of this equation can be used for ephemeral gully prediction (see below).

The 'sediment transporting capacity index', developed by Moore *et al.* (1992) is very similar to the simple stream power index. This index is also a function of local slope and contributing area

$$\tau = \left[\frac{A_s}{22 \cdot 13}\right]^{0.6} \cdot \left[\frac{\sin \beta}{0 \cdot 0896}\right]^{1 \cdot 3} \tag{5}$$

where τ = sediment transport index, A_s = unit contributing area (m^2/m) and β = slope angle (degrees).

A map of the sediment transport capacity index for the Catsop catchment is shown as Figure 8. Although there are differences, the map is very similar to that of the LS-factor using contributing area (Figure 4).

Desmet and Govers (1996) also developed a sediment transporting capacity index based on a soil erosion potential index. This index can be used for predicting ephemeral gully initiation (Desmet *et al.*, 1998). For

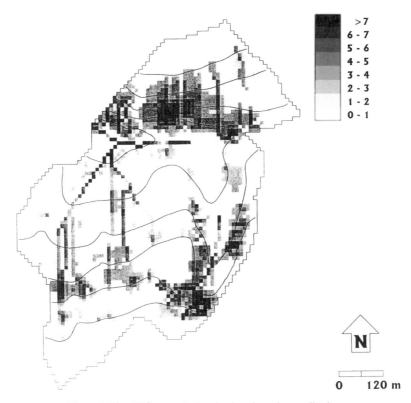

Figure 4. The *LS*-factor calculated using the unit contributing area

this purpose, only contributing area with arable land use should be counted. All grassland fields have to be omitted. Their equation is

$$T_c = k_2 \cdot k_1 \cdot S^m \cdot A_s^n \tag{6}$$

where T_c = transporting capacity of the overland flow (mass/unit length of contour), k_2 = a proportionality factor (length), k_1 = coefficient of proportionality (constant, here 4), S = slope gradient (m/m), m = slope exponent (constant, here 1·4), A_s = contributing area per unit of contour length (m^2/m) and n = area exponent (constant, here 0·5). Figure 9 shows a map of this sediment transporting capacity index for the Catsop catchment. For ease of calculation and to show the similarities with Figures 7 and 8, grassland is included here, not omitted.

Combining the above-mentioned indices with soil erodibility and crop cover information, it is possible to produce very powerful erosion hazard indicators. An example of such a combined index is the soil erosion hazard index developed by Vertssy *et al.* (1990)

$$H = \rho \cdot g \cdot A_s \cdot i \cdot S \cdot \alpha_1 \cdot e^{-\alpha_2 \cdot (1 - C_r)} \tag{7}$$

where H = erosion hazard index, ρ = density of water, g = acceleration due to gravity, A_s = contributing area, i = rainfall excess rate, S = slope gradient (m/m), α_1 = constant depending on soil type and surface management (0·05–0·7), α_2 = constant depending on soil type and surface management and $1 - C_r$ = the fraction of surface cover in contact with the ground.

Study of Figure 10 demonstrates that the soil erosion hazard index is mainly concerned with rill and gully initiation risk, since there is a strong similarity with the concentrated flow lines on arable land: locations with

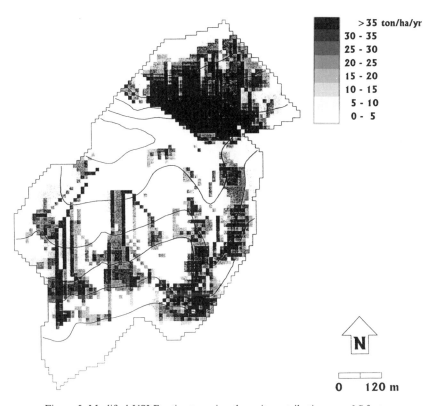

Figure 5. Modified USLE estimates using the unit contributing area *LS*-factor

a large upstream area. Upstream area seems to be the most sensitive variable in this equation. Splash and interrill erosion appear to be of minor importance.

Automatic calculation of aspect, flow direction or local drainage direction are not successful whenever local factors are not included. For example Ludwig *et al.* (1996) showed the influence of tillage direction, wheel tracks and local linear features at field boundaries. Because of tillage, the 'normal' topographic flow direction and flow path can be changed and forced into the direction of tillage. This depends on the angle between the topographical flow direction and the tillage direction, and on the soil surface roughness. They showed that tillage and other agricultural effects lead to very different overland flow patterns. Thus, peak flows and local erosion estimations can be very different from one model to another, depending on whether or not these factors are taken into account. Therefore, it is often necessary to modify the drainage pattern calculated by the GIS manually.

Another improvement in soil erosion estimation can be achieved if remotely sensed land use data are used, especially when large areas are to be simulated. For example, De Jong (1994) used remote sensing data and GIS to estimate soil erosion amounts in large areas in the Ardèche (France). Using this approach, high spatial resolution can be obtained, including within-field variability of crop cover. Future remote sensing techniques might also give high resolution information on soil properties.

LOOSE COUPLING OF GIS AND RUNOFF AND EROSION MODELS

In the previous section it has been shown that standard GIS techniques can be used to produce soil erosion hazard indices and quantitative estimates based on USLE-type calculations. However, more advanced models are available. Some of these models have been loosely coupled to a GIS, using conversion programs

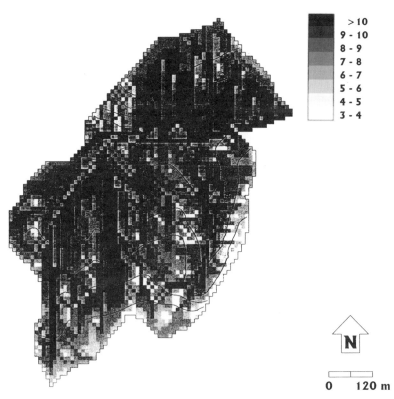

Figure 6. The wetness index in the Catsop catchment

or user interfaces. To couple the models and the GIS the output maps of a GIS are converted to the specific input format needed by the model. The model is run, and the output from the model is converted to the input format needed to import an erosion map, for example, back into the GIS, properly. The advantage of this procedure is that the model need not be changed. The conversion program must however be changed whenever a new GIS format is used. The main disadvantage is that the procedure is time-consuming, and model input files become extremely large. Also, errors in the model input file are difficult to trace. Examples of erosion models that have been linked to a GIS are the ANSWERS model (De Roo *et al.*, 1989) and the AGNPS model (Mitchell *et al.*, 1992).

The ANSWERS (<u>a</u>real <u>n</u>on-point <u>s</u>ource <u>w</u>atershed <u>e</u>nvironment <u>r</u>esponse <u>s</u>imulation) distributed parameter model, developed by Beasley (Beasley and Huggins, 1982), used for modelling surface runoff and soil erosion, has been fully integrated within a GIS by writing interfacing programs to input data from the GIS to the model and to display the results (De Roo *et al.*, 1989). The catchment to be modelled is assumed to be composed of square elements. Values of key variables are defined for each element, e.g. slope, aspect, soil variables (porosity, moisture content, field capacity, infiltration capacity, erodibility factor), crop variables (coverage, interception capacity, USLE *C/P*-factor), surface variables (roughness and surface retention) and channel variables (width and roughness). A rainfall event is simulated with time increments of one minute, taking into account spatial and temporal variability of rainfall. Runoff and soil erosion processes and their associated parameters vary over the catchment. If data about these parameters are stored in geographical information systems they can easily be extracted for this type of erosion modelling. It has been shown by De Roo *et al.* (1989) that using the GIS for modelling changes the model results, when compared with a more lumped approach to modelling. For a rainfall event in 1987, total runoff was 46% higher for the distributed case (4275 units/grids) than for the lumped case, where only 20 units were used; the soil loss was 36% higher, and peak runoff 42% greater. From this it is clear that distributed parameter models such as ANSWERS

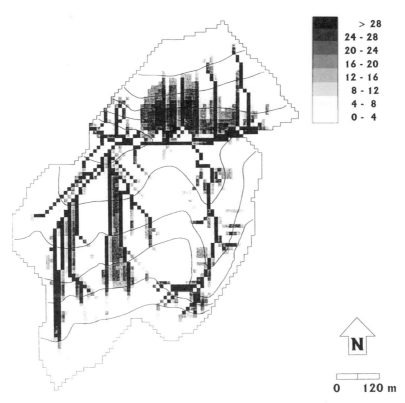

> 28
24 - 28
20 - 24
16 - 20
12 - 16
8 - 12
4 - 8
0 - 4

N

0   120 m

Figure 7. The stream power index in the Catsop catchment

cannot be optimally utilized without the use of a GIS to supply data at the correct spatial resolution. Further advantages of using a GIS are the possibilities of producing and evaluating land use scenarios, producing spatial and temporal patterns of both erosion/deposition (Figure 11) and overland flow. These maps can be compared by subtraction to yield maps indicating how erosion or sedimentation might be affected by certain control measures within the catchment.

EMBEDDED COUPLING OF GIS AND RUNOFF AND EROSION MODELS

One step further is the full integration of models in a GIS, which is also called embedded coupling (Wesseling *et al.*, 1995). Using this approach, conversion of input and output data are no longer necessary. There are two types of embedded coupling. The first type is adding a simple GIS to a complex modelling system to display results and provide interactive control. This approach has been used by Schmidt (Schmidt, 1991; Von Werner and Schmidt, 1997) in the development of EROSION3D model. The second type of embedded coupling is where the model is written using the analytical engine of the GIS. The great advantage of embedded coupling is that the user can construct his or her own GIS-based model as required, without having to write routines for displaying data and other standard GIS operations, and without writing large conversion programs. The disadvantage of embedded coupling is that the analytical power of the model is constrained by the power of the user interface. Complex mathematical modelling is often not possible in current GIS (Wesseling *et al.*, 1995). Sophisticated use of embedded models in most commercial GIS is currently not easy, because the GIS are not designed to deal specifically with spatiotemporal processes, or processes in which mass balance and mass transport are integral features. The PCRaster dynamic modelling GIS (Wesseling *et al.*, 1995) is a GIS together with a programming language such that complex models can

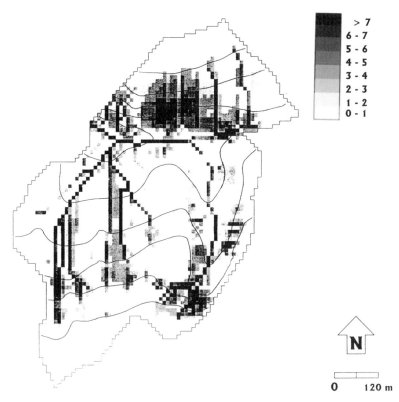

	> 7
	6 - 7
	5 - 6
	4 - 5
	3 - 4
	2 - 3
	1 - 2
	0 - 1

N

0   120 m

Figure 8. The sediment transport capacity index (Moore *et al.*, 1992) in the Catsop catchment

be developed. The runoff and erosion model LISEM (De Roo *et al.*, 1996a) is an example of an embedded model in the PCRaster GIS.

The Limburg soil erosion model (LISEM) is a physically based hydrological and soil erosion model developed by the Department of Physical Geography of Utrecht University and the Soil Physics Division of the Winand Staring Centre in Wageningen, The Netherlands. The model simulates runoff and erosion in catchments during and immediately after a single rainfall event. The model can be used for scientific, planning and conservation purposes. Processes incorporated into the model, which are described below, are rainfall, interception, surface storage in microdepressions, infiltration, vertical movement of water in the soil, overland flow, channel flow, detachment by rainfall and throughfall, detachment by overland flow and the transport capacity of the flow. Also, the influence of tractor wheelings and small paved roads (smaller than the pixel size) on the hydrological and soil erosion processes are taken into account. A flowchart of LISEM is presented in Figure 12. An extensive description of the model can be found in De Roo *et al.* (1996a). Figures 13 and 14 give examples of GIS output of LISEM for the Catsop catchment. For every user-defined time increment, runoff maps can be produced. Also, a summary output file with totals and hydrographs and sedigraphs are produced for the outlet and for two user-defined suboutlets within the catchment.

The big advantage of an embedded model is that the 'modeller' can concentrate on the basic model development, i.e. entering and changing equations to describe the hydrological and erosion processes. The structure of the modelling language obviates the need for the modeller to programme database structures, display tools or model optimizing algorithms. This is all taken care of by the modelling language. Using LISEM provides a convenient way of simulating different scenarios in a catchment, as demonstrated by De Roo *et al.* (1996b).

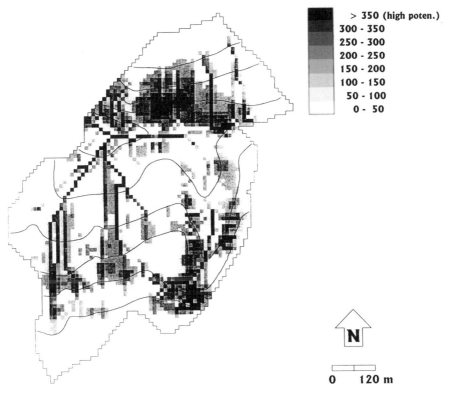

Figure 9. The sediment transporting capacity index (Desmet and Govers, 1996) in the Catsop catchment. Grassland is not omitted here

USING GIS IN ANALYSING VALIDATION PROBLEMS

Apart from employing GIS as an easy-to-use database and toolkit for modelling, other GIS techniques can be used to assess the uncertainty and validity of spatial erosion models. In this section three GIS-related topics will be addressed: (1) validating distributed soil erosion models with ^{137}Cs-derived point estimates of actual soil loss; (2) using spatial Monte Carlo simulation to assess the changes in model output due to uncertainty in infiltration variables; and (3) using PCRaster GIS and Monte Carlo simulation to assess the changes in runoff pathways due to uncertainty in digital elevation data.

To examine the erosion problem and to validate distributed models that can predict soil erosion, spatially distributed data on actual soil loss are needed. [137]Cs measurements can be used to provide spatially distributed point estimates of soil loss. De Roo (1991, 1993) and De Roo and Walling (1994) demonstrated that these ^{137}Cs-derived soil erosion estimates can be compared to simulated soil loss. The results showed that the relatively simple soil erosion models USLE and the Morgan/Morgan/Finney model (MMF) (Morgan *et al.*, 1984) are little worse than the more complex physically based soil erosion models KINEROS (Woolhiser *et al.*, 1990) and ANSWERS (Beasley and Huggins, 1982). The correlations with the measured ^{137}Cs data were similar for the four models examined. The correlations might improve if the recently demonstrated influence of soil tillage on soil redistribution (Govers *et al.*, 1993; Quine *et al.*, 1994) is taken into account.

There are many reasons why simulation results produced by distributed erosion models are often different from the observed patterns (De Roo, 1993, 1996). There can be uncertainties in the theoretical structure of the model, errors in the mathematical solution methods in the model, calibration errors and errors and uncertainties in the data. Given the fact that, in general, only limited data are available for validation,

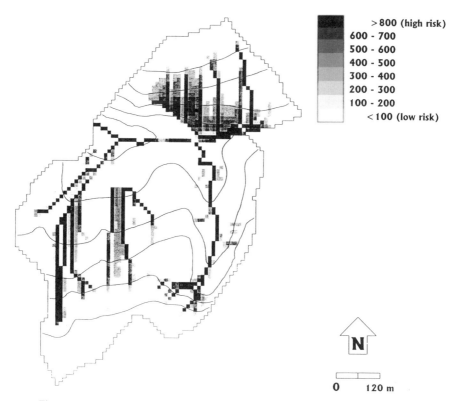

Figure 10. The soil erosion hazard index (Vertessy *et al.*, 1990) in the Catsop catchment

e.g. only measurements of sediment output, a meaningful assessment of physically based models is very difficult. One of the uncertainties in input data is the uncertainty in a DEM. A DEM is often created by interpolation error is an important source of uncertainty. Furthermore, tillage effects can lead to minor (local) changes in altitude. Normally, a single DEM is used for modelling, leading to a single result. If the uncertainty is taken into account by undertaking, for example, 100 replicate simulations involving adding a random error to the standard DEM, 100 different runoff pathways are produced. Figure 15 shows the probability of a pixel being part of the flow network after 100 simulations using a 5 cm standard deviation in the DEM. A pixel is classified as a member of the flow network if the contributing area is 1 hectare. The uncertainty in runoff routing leads to uncertainty in both overland flow rate and channel flow rate and therefore to uncertainty in the erosion estimates.

A similar GIS Monte Carlo (MC) approach was used to assess the effects of uncertainty in infiltration variables on the output of the ANSWERS model (De Roo *et al.*, 1992). Normally, an individual agricultural field would be assigned a fixed uniform infiltration value or some spatial pattern of infiltration values derived from interpolation. In the MC procedure, the standard deviation in the measurements has been used. In this way, instead of a single hydrograph, 50 or 100 hydrographs are produced, leading to an error band around a mean hydrograph.

WHAT DO WE GAIN WHEN USING DETAILED SPATIALLY DISTRIBUTED DATA?

In theory, detailed information on slope profile and land use incorporated in a GIS-based model should lead to better results than when a lumped approach is used for soil erosion modelling. However, at present,

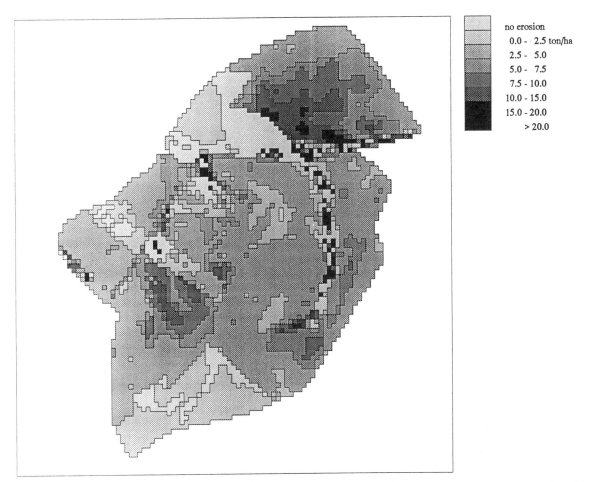

Figure 11. Total soil erosion and deposition in the Catsop catchment between 1987 and 1990, simulated using the ANSWERS model: a summation of events

detailed physically based erosion models do not necessarily produce better results than much simpler and partly lumped erosion models with 'representative elements', although they reproduce topography in more detail. This was demonstrated by De Roo (1993) for soil erosion models, and was also one of the conclusions of a GCTE Soil Erosion Network Conference held in Utrecht in 1997 (Jetten and De Roo, 1998), during which simulation results from several erosion models were compared with measured water and sediment discharge. This finding is also a common conclusion for catchment hydrological models more generally. Another interesting conclusion of the GCTE conference was the fact that any USLE-based approach to modelling appears to be weak in simulating extreme events: sediment discharge (scouring, channel processes) is underpredicted.

The reasons for the disappointing results obtained from spatially distributed models must be sought in the uncertainty involved in estimating and measuring the large number of input variables at a catchment scale. There is a need for much simpler models that use readily available and significant spatial data, such as topography and land use, and only use a very limited number of other key variables, such as infiltration capacity, soil texture and initial soil moisture content. In such models only the dominant processes should be simulated, to prevent the model from being overparameterized. The GIS techniques used in the first part of this paper could be helpful for the development of these simple models.

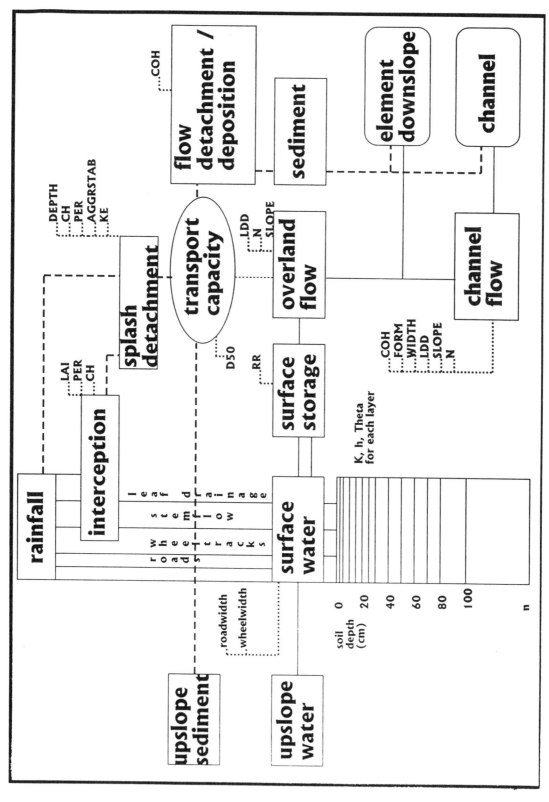

Figure 12. Flowchart of the LISEM model

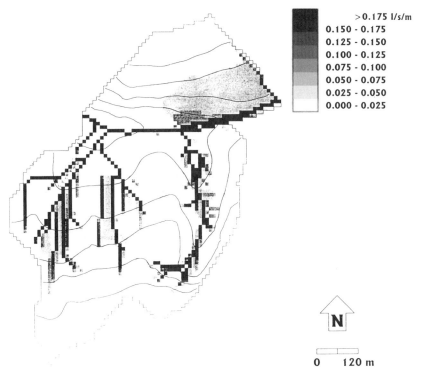

Figure 13. Surface runoff in the Catsop catchment on 30 May 1993 at 1555 hours, simulated with the LISEM model

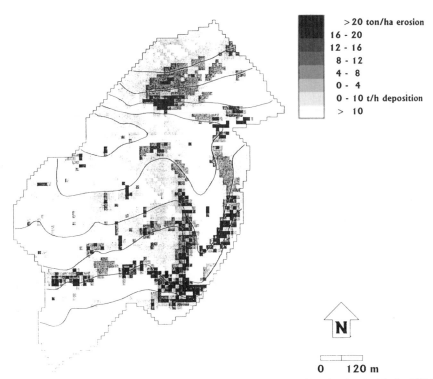

Figure 14. Erosion and deposition in the Catsop catchment on 30 May 1993, simulated with the LISEM model

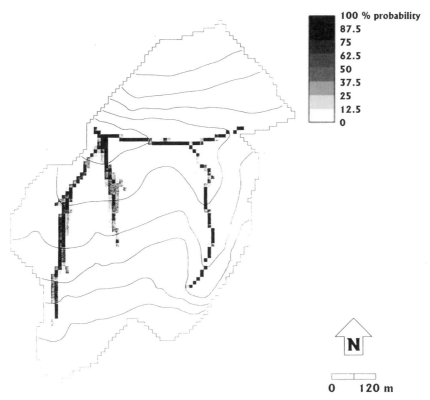

Figure 15. The effects of a 5 cm uncertainty in the DEM of the catsop catchment on the 'membership' of pixels to the flow network

DISCUSSION AND CONCLUSIONS

The use of geographical information systems in catchment hydrological and erosion modelling offers considerable potential. There is a clear distinction between erosion hazard modelling, which is primarily concerned with estimating on-site erosion rates, and the erosion and sediment yield modelling. In theory, the latter models could be lumped and do without GIS. The on-site erosion prediction models clearly need GIS. However, even the yield models benefit greatly from using a GIS. Many studies have shown the importance of including spatial variability of, for example, rainfall, infiltration and slope gradient.

Topography and land use can be simulated in great detail and data management and presentation is much easier than when a GIS is not used. Many powerful methods exist to derive useful products from a DEM such as slope gradient, flow direction, contributing area and slope length. From these basic variables, a wide range of erosion estimates can be derived. These variables can also be used in more complex physically based models either linked to, or integrated in, a GIS. However, GIS raster-based erosion models do not necessarily produce better results than much simpler and partly lumped erosion models with 'representative elements', although they reproduce topography in more detail. Reasons for the disappointing results of spatial models must be sought in the uncertainty involved in estimating and measuring the large number of input variables at a catchment scale. The uncertainty is sometimes summed such that the predictions are even worse than lumped models, and sometimes the uncertainties are balanced out such that the model does predict realistic sediment yields. There is a need for much simpler models simulating only the dominant process operating in the catchment. Many scientists still concentrate on the development of the theoretical structure of the model and try to fill the gaps in our understanding of hydrological and soil erosion processes.

This is very important work, but the errors in the data may have a larger effect on the simulation results obtained with the model.

ACKNOWLEDGEMENTS

My former colleagues at the Department of Physical Geography at Utrecht University are thanked for all their stimulating ideas, remarks, GIS programs and contributions to GIS model development over the years. They include: Prof. Dr P. A. Burrough, Dr S. M. De Jong, Mr L. Hazelhoff, Mr C. G. Wesseling, Dr V. G. Jetten, Dr W. P. A. Van Deursen, Mr E. J. Henkens, Mr J. C. Van Hees, Dr H. Th. Riezebos and Dr Th. W. A. Van Asch.

REFERENCES

Amarú, M., Van Dijck, S. J. E., Van Asch, Th. W. A. 1996. 'Predicting the impact of land use change on event-based runoff in a small Mediterranean catchment', *ITC J.*, **1995-4**, 331–336.

Beasley, D. B. and Huggins, L. F. 1982. *ANSWERS — User's manual*. Dept. of Agr. Eng., Purdue University, West Lafayette, In.

Beven, K. J. and Kirkby, M. J. 1979. 'A physically based, variable contributing area model of basin hydrology', *Hydrol. Sci. Bul.*, **24**, 43–69.

Boardman, J., Ligneau, L., De Roo, A. P. J. and Vandaele, K. 1994. 'Flooding of property by runoff from agricultural land in northwestern Europe', *Geomorphology*, **10**, 183–196.

Bork, H. R. and Hensel, H. 1988. 'Computer-aided construction of soil erosion and deposition maps', *Geol. Jahr.*, **A104**, 357–371.

Burrough, P. A. 1986. *Principles of Geographical Information Systems for Land Resources Assessment*. Clarendon Press, Oxford. 193 pp.

De Jong, S. M. 1994. 'An erosion model integrating remotely sensed data with GIS data: SEMMED', *Applications of Reflective Remote Sensing for Land Degradation Studies in a Mediterranean Environment, Netherlands Geographical Studies*, **177**. University of Utrecht, Utrecht. pp 75–94.

De Roo, A. P. J. 1991. 'The use of ^{137}Cs as a tracer in an erosion study in South-Limburg (The Netherlands) and the influence of Chernobyl fallout', *Hydrol. Process*, **5**, 215–227.

De Roo, A. P. J. 1993. *Modelling surface runoff and soil erosion in catchments using geographical information systems; Validity and applicability of the 'ANSWERS' model in two catchments in the loess area of South Limburg (The Netherlands) and one in Devon (UK), Netherlands Geographical Studies*, **157**. University of Utrecht, Utrecht. 304 pp.

De Roo, A. P. J. 1996. 'Validation problems of hydrologic and soil erosion catchment models: examples from a Dutch erosion project', in Anderson, M. G. and Brooks, S. (eds), *Advances in Hillslope Processes*. John Wiley & Sons. pp. 669–683.

De Roo, A. P. J. and Walling, D. E. 1994. 'Validating the 'ANSWERS' soil erosion model using 137Cs', in Rickson, R. J. (ed.), *Conserving Soil Resources. European Perspectives*. CAB International, Cambridge. pp. 246–263.

De Roo, A. P. J., Eppink, L., Hazelhoff, L., Jessen, G., Burrough, P. A. and Schouten, C. J. 1987. 'Onderzoek wateroverlast en bodemoerosie nabij Catsop (gem. Stein, Limburg', *De Landinrichtingsdienst, het Staatsbosbeheer, de provincie Limburg en het Waterschap Roer en Overmaas*. Rijksuniversiteit Utrecht/Landbouwuniversiteit Wageningen. Roermond. 109 pp.

De Roo, A. P., Hazelhoff, L. and Burrough, P. A. 1989. 'Soil erosion modelling using 'ANSWERS' and geographical information systems', *Earth Surf. Proces. Landf.*, **14**, 517–532.

De Roo, A. P. J., Hazelhoff, L. and Heuvelink, G. B. M. 1992. 'The use of Monte Carlo simulations to estimate the effects of spatial variability of infiltration on the output of a distributed hydrological and erosion model', *Hydrol. Process.*, **6**, 127–143.

De Roo, A. P. J., Wesseling, C. G. and Ritsema, C. J. 1996a. 'LISEM: a single event physically based hydrological and soil erosion model for drainage basins: I. Theory, input and output', *Hydrol. Process.*, **10**, 1107–1117.

De Roo, A. P. J, Offermans, R. J. E. and Cremers, N. H. D. T. 1996b. 'LISEM: a single event physically based hydrological and soil erosion model for drainage basins: II. Sensitivity analysis, validation and application', *Hydrol. Process.*, **10**, 1119–1126.

Desmet, P. J. J. and Govers, G. 1995. 'GIS-based simulation of erosion and deposition patterns in an agricultural landscape: a comparison of model results with soil map information', *Catena*, **25**, 389–401.

Desmet, P. J. J. and Govers, G. 1996. 'A GIS procedure for automatically calculating the USLE LS factor on topographically complex landscape units', *J. Soil Wat. Conserv.*, **515**, 427–433.

Desmet, P. J. J., Poesen, J. and Govers, G. 1997. 'Relative importance of slope gradient and contributing area for optimal prediction of the initiation and trajectory of ephemeral gullies', *Catena*, in press.

Dräyer, D. 1996. 'GIS-gestützte Bodenerosionsmodellierung im Nordwestschweizerischen TafejJura — Erosionsschadenskartierungen und Modellergebnisse', *Basler Beitrage zur Physiographie: Physiogeographica*, Vol. **22**. 234 pp.

Ferri, V. and Minacapilli, M. 1995. 'Sediment delivery processes at the basin scale', *Hydrol. Sci. J.*, **40**, 703–717.

Fröhlich, J., Dräyer, D. and Huber, M. 1994. 'GIS-Methoden in der landschaftsökologischen Raumbewertung mit einem Beispiel zur Bestimmung der Bodenerosionsgefährdung', *Die Erde*, **125**, 1–13.

Govers, G., Quine, T. A., and Walling, D. E. 1993. 'The effects of water erosion and tillage movement on hillslope profile development: a comparison of field observations and model results', in Wicherek, S. (ed) *Farm Land Erosion in Temperate Plains and Hills*. Elsevier, Amsterdam. pp. 285–300.

Jetten, V., Henkens, E. J., De Jong, S. M. 1988. *The Universal Soil Loss Equation. Version 1·0, release 1·0, distributed.* Department of Physical Geography, Utrecht University, The Netherlands.

Jetten, V., De Roo, A. P. J., and Lorentz, S. 1998. Validating soil erosion models at the catchment scale. *CATENA.* In press.

Ludwig, B., Daroussin, J., King, D. and Souchere, V. 1996. 'Using GIS to predict concentrated flow erosion in cultivated catchments', *IAHS Publ.,* **235**, 429–436.

Mitchell, J. K., Engel, B. A., Srinivason, R. and Wang, S. S. Y. 1992. 'Validation of AGNPS for small watersheds using an integrated AGNPS/GIS system', *Paper no. 92-2532.* ASAE. 14 pp.

Moore, I. D., Gessler, P. E., Nielsen, G. A. and Peterson, G. A. 1992. 'Terrain analysis for soil-specific crop management', in *Soil Specific Crop Management: A Workshop on Research and Development Issues.* Minnesota Extension Service, University of Minnesota (Agriculture), Minneapolis. 23 pp.

Moore, I. D., Turner, A. K., Wilson, J. P., Jenson, S. K. and Band, L. E. 1993. 'GIS and land–surface–subsurface process modelling', in Goodchild, M. F., Parks, B. O. and Steyaert, L. T. (eds), *Environmental modelling with GIS.* pp. 213–230.

Morgan, R. P. C., Morgan, D. D. V. and Finney, H. J. 1984. 'A predictive model for the assessment of soil erosion risk', *J. Agric. Engng. Res.,* **30**, 235–253.

Quine, T. A., Desmet, P. J. J., Govers, G., Vandaele, K., and Walling, D. E. 1994. 'A comparison of the roles of tillage and water erosion in landform development and sediment export on agricultural land near Leuven, Belgium', *IAHS Publ.,* **224**, 77–86.

Schmidt, J. 1991. 'A mathematical model to simulate rainfall erosion', *Catena,* **19** (Suppl.), 101–109.

Van Hees, J. C., Henkens, E. J., De Jong, S. M. and De Roo, A. P. J. 1987. 'A land evaluation and soil erosion study in the Ardeche', *Msc Thesis.* Internal Report, Utrecht University.

Vertessy, R. A., Wilson, C. J., Silburn, D. M., Connolly, R. D. and Ciesiolka, C. A. 1990. 'Predicting erosion hazard areas using digital terrain analysis', *Proceedings IAHS International Symposium on Research Needs and Applications to Reduce Erosion and Sedimentation in Tropical Steeplands, Suva, Fiji, 11–15 June 1990.*

Von Werner, M. and Schmidt, J. 1997. 'EROSION 2D/3D — Ein Computermodell zur Simulation der Bodenerosion durch Wasser, Band III: EROSION 3D — Modellgrundlagen, Bedienungsanleitung. Hrsg.: Sächsische Landesanstalt für Landwirtschaft, Sächsisches Landesmt für Umwelt und Geologie.

Wainwright, J. 1996. 'Hillslope response to extreme storm events: the example of the Vaison-La-Romaine event', in Anderson, M. G. and Brooks, S. (eds), *Advances in Hillslope Processes.* John Wiley & Sons. pp. 997–1026.

Wesseling, C. G., Karssenberg, D., Van Deursen, W. P. A. and Burrough, P. A. 1995. 'Integrating dynamic environmental models in GIS: the development of a dynamic modelling language', *Trans. GIS,* **1**, 40–48.

Woolhiser, D. A., Smith, R. E. and Goodrich, D. C. 1990. 'KINEROS: a kinematic runoff and erosion model: documentation and user manual', *USDA — ARS, ARS-77.* USDA–ARS.

8

DECIPHERING LARGE LANDSLIDES: LINKING HYDROLOGICAL, GROUNDWATER AND SLOPE STABILITY MODELS THROUGH GIS

DANIEL J. MILLER[1,2]* AND JOAN SIAS[2]

[1] *M2 Environmental Services, 4232 5th Ave. NW, Seattle, WA 98107, USA*
[2] *Earth Systems Institute, 1314 NE 43 St., Suite 207, Seattle, WA 98105, USA*

ABSTRACT

Large landslides can deliver substantial volumes of sediment to river channels, with potentially adverse consequences for water quality and fish habitat. When planning land use activities, it is important both to consider the risks posed by landslides and to account for the effects of land use on rates of landslide movement. Of particular interest in the Pacific Northwest are the effects of timber harvest in groundwater recharge areas of landslides. Because of variability between sites, and variability over time in precipitation and other natural environmental factors affecting landslide behaviour, empirical data are usually insufficient for making such determinations. We describe here the use of simple numerical models of site hydrology, groundwater flow and slope stability for estimating the effects of timber harvest on the stability of the Hazel Landslide in northwestern Washington State. These effects are examined relative to those of river bank erosion at the landslide toe. The data used are distributed in time and space, as are the model results. A geographical information system (GIS) provides an efficient framework for data storage, transfer and display. Coupled with process-based numerical models, a GIS provides an effective tool for site-specific analysis of landslide behaviour.

KEY WORDS   landslides; GIS; hydrology; groundwater; slope stability; forest hydrology

INTRODUCTION

Large landslides involve a broad range of mass-wasting processes. Even a single landslide, or landslide complex, can exhibit a variety of behaviours and respond to a variety of environmental factors, some of which may be removed from the landslide itself. Factors affecting groundwater flow, for example, may be distributed over a large area. These issues present a recurring challenge to natural resource managers in areas of active or potential landslide activity. Simple avoidance of landslide-prone areas can be an expensive option, and may not be necessary if land use activities will have no influence on landslide behaviour.

Characterizing the effects of distributed land use on landslide activity is traditionally an empirical endeavour (e.g. Sidle *et al.*, 1985). To assess the effects of forest management on landslide occurrence, for example, Washington State has implemented a procedure (WFPB, 1995) that relies on correlations between landslide rates (e.g. number of landslides per square kilometre per year) and forestry activities (e.g. timber harvest and road construction). This procedure works satisfactorily for small, shallow landslides. Unfortunately, large landslide complexes are not as amenable to empirical categorization. For example, timber harvest can increase groundwater levels over a period of years (see Hammond *et al.*, 1992 for a review) with consequent increases in rates of mass movement (Swanston *et al.*, 1988). Cause and consequence may be separated in both space and time. The diversity of the interacting factors potentially affecting

---

* Correspondence to: Daniel J. Miller, M2 Environmental Services, 4232 5th Ave NW, Seattle, WA 98107, USA.

the behaviour of large landslides confounds attempts to find simple relationships between land use and landslide activity.

Careful monitoring may reveal apparent connections between forestry activities and mass movement rates (Swanston *et al.*, 1988). However, monitoring programmes require funding, perhaps over the course of many years. Extended funding is difficult to obtain without some indication that the costs will eventually pay off. Moreover, to find the link between forest land management and landslide activity, or a lack thereof, requires that some forestry activity is actually performed, an experiment that may not be allowed if the potential consequences of accelerated mass wasting are considered too dire.

An alternative is first to simulate the experiment numerically, a feat that has become feasible only with the advent of geographic information systems (GIS). With GIS, simple process-based models using digital data can be applied over broad areas (Dietrich *et al.*, 1995; Miller, 1995). Various scenarios can be examined, with parameter values altered to reflect a proposed land use. The results offer an estimate of landslide response with which to evaluate the risk of actually performing the land use experiment, and can guide efforts at data collection and monitoring.

These methods can also be used to formulate hypotheses to test and improve our understanding of the processes active within a large landslide. Model response to a change in any of the pertinent variables (groundwater recharge, for example) can isolate the role of various factors that affect slope stability. Spatially distributed data allow explicit prediction of the spatial pattern of slope response to such changes. By simulating an historical change in any of the driving variables, we predict a landslide response to compare with what has been observed.

The key to this approach is calculation of a response to a change in some controlling variable. We find that attempts to predict slope stability directly, i.e. a factor of safety, meet with less success. The reason is that the data available for such broad application of a slope stability model are very limited. We lack, for example, any geotechnical tests, so that rock and soil property values are constrained only by the range of values reported for similar rock and soil types. The change in the factor of safety calculated for a change in any model variable is a more robust predictor of slope sensitivity to environmental perturbations than the 'factor of safety' values themselves.

Here, we illustrate the use of process-based models implemented through a GIS with an analysis of the Hazel Landslide, a large landslide complex in glacio lacustrine sediments adjacent to the Stillaguamish River in north-western Washington, USA. Sediment shed from this landslide has posed a chronic threat to an important fishery for decades. Upslope areas are managed for timber production. This study was motivated by concerns that timber harvesting within the groundwater recharge area of the landslide could increase groundwater flow, and consequently increase sediment flux from the landslide. Previous studies have highlighted the potential importance of groundwater flow to the landslide (Shannon and Associates, 1952; Benda *et al.*, 1988), but were unable to quantify its role relative to other factors (WDNR, 1996).

For this analysis, we were asked to estimate the potential effects of clear-cut timber harvesting in areas upslope of the landslide (no harvests are proposed on the landslide itself) relative to the effects of bank erosion of the landslide toe. The data available were: a 1 : 4800-scale topographic map of the area; time-series of precipitation and temperature from a regional weather station; and field estimates of stratigraphic contact and surface seep locations. The stratigraphy consists of a thick sequence of glaciofluvial sediments (primarily sand) overlaying lacustrine silts and clays. Surface seeps are found year-round, emerging at the sand–clay contact in many places. Some of these seeps are associated with areas of activity on the Hazel Landslide. The aerial distribution and flux from the seeps were mapped in autumn (the end of the dry season) and spring (the end of the wet season). The topographic and field-mapped data were digitized and used with simple models to estimate: (1) recharge to groundwater; (2) the aerial pattern of groundwater flow (assuming the outwash deposits act as an unconfined aquifer); and (3) slope stability over the area of the landslide. Data transfer and display were accomplished within a GIS. We used the calculated water table to estimate the extent of the groundwater recharge area to the landslide and then examined model responses to: (1) clear-cut timber harvesting of the groundwater recharge area; (2) bank erosion of the landslide toe, and; (3) incision of small stream channels draining the body of the landslide. The results indicate large spatial variability in the magnitude of the response to each of these changes, with patterns of response unique to each process. Compared with patterns of landslide activity observed in historical sequences of aerial photographs and in the field, these results offered hypotheses to explain landslide behaviour and provided insight that guided land management decisions.

Our modelling strategy was dictated in part by the characteristics of the Hazel Landslide. Although the techniques presented here have much broader applicability, it is useful to describe them in the context of their application in this case, so we start below with a description of the site.

Site characteristics

The Hazel Landslide sits at the south-east end of the Whitman Bench (Figure 1), a large terrace remnant composed of lacustrine clays underlying glaciofluvial outwash, a stratigraphic sequence common in valleys draining the Cascade Mountains west to Puget Sound (Tabor *et al.*, 1988; Booth, 1989). Extensive fluvial incision of these deposits following the Vashon stade of the Fraser Glaciation (12 000 years) created conditions conducive to mass wasting. Morphology indicative of large, deep-seated landsliding within these deposits is common throughout this region. Although most such landslides appear to be dormant, several recently and currently active deep-seated landslides, such as the Hazel Landslide, attest to the continuing potential for mass movement of this material (e.g. Thorsen, 1989). Several large, post-glacial slump blocks (currently dormant) scallop the margins of the Whitman Bench. As shown in Figure 1, the Hazel Landslide is contained within one of these blocks and the surrounding topography indicates adjacent slumps to the west and to the north.

The climate at Hazel is maritime, with cool, humid winters and dry summers. Annual average rainfall at the nearest weather station (the US Forest Service ranger station in Darrington, Washington; 17·6 kilometres to the east) is 2250 mm per year (1931–1994), with 84% of rainfall occurring from October to April. Average January and July air temperatures are 2·1 °C (36 °F) and 19·5 °C (67 °F), respectively. We assume 24-hour precipitation at Hazel is 78% that at Darrington (Wolff *et al.*, 1989) and that there is equivalence of air temperature at the two sites. Digital meteorological data were available for the period 1931–1994.

In addition to geographical proximity, there are other relevant similarities between Hazel and Darrington that further justify the suitability of these data. These include: elevation (the climate station is located at 500 feet; the toe of the landslide and the Whitman Bench are at elevations of about 300 and 800 feet, respectively); aspect (both are oriented to the south-west); exposure to frontal weather; and characteristics of upwind terrain — with respect to both topography and vegetation. On the other hand, the meteorological data are collected over a clearing, and therefore may not be very representative of meteorological conditions over a forest, quite possibly introducing a bias towards overestimation of forest (but not clear-cut) evapotranspiration (Pearce *et al.*, 1980).

The Hazel Landslide exhibits several types of activity (Figure 2; Shannon and Associates, 1952; Benda *et al.*, 1988; Miller and Sias, 1997). The most spectacular are slumps that occur in steep scarps along the river edge and evolve into debris avalanches (following the classification of Varnes, 1978). These have at times temporarily dammed the Stillaguamish River and diverted the channel considerably southward. These riverside events at the toe of the landslide deliver a considerable volume of fine material to the river channel and create bank deposits that erode rapidly. Several small, groundwater-fed streams drain the body of the landslide. Slumps along these stream channels form a chronic source of fine sediment to the Stillaguamish River (Shannon and Associates, 1952). Landslide activity extends a considerable distance upslope from the river edge and involves multiple blocks in various stages of evolution. Headward growth typically occurs via a series of rotational and translational slumps that expose steep head scarps in the outwash sands and have slip surfaces extending into the underlying lacustrine deposits. These slumps typically evolve downslope into mud flows. Movement tends to be slow, although rapid translation of large slumps has also occurred. Rates of movement vary over the landslide and tend to be punctuated, with the locus of activity shifting in space and time.

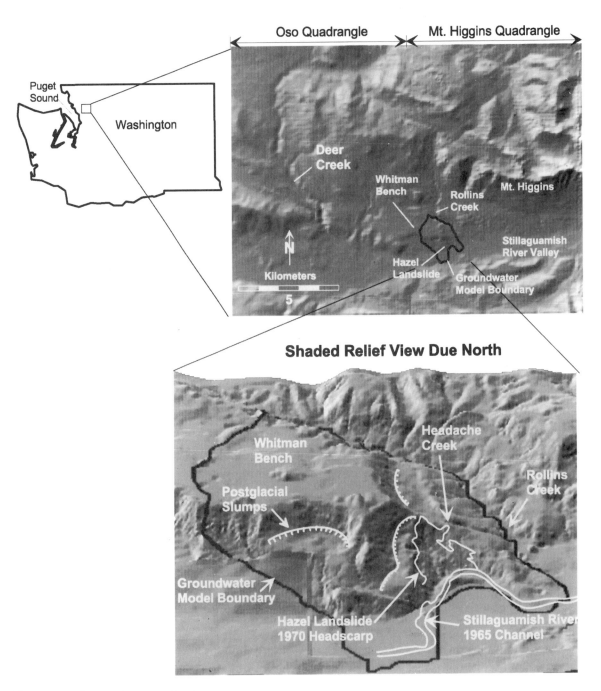

Figure 1. Location and site topography of the Hazel Landslide. The upper, shaded-relief image is created from US Geological Survey 30-m grid DEMs for the Oso and Mt Higgins 7½ minute quadrangles; the lower image is created from a 10-m grid DEM interpolated from 20-foot contour lines on a 1:4800-scale topographic map

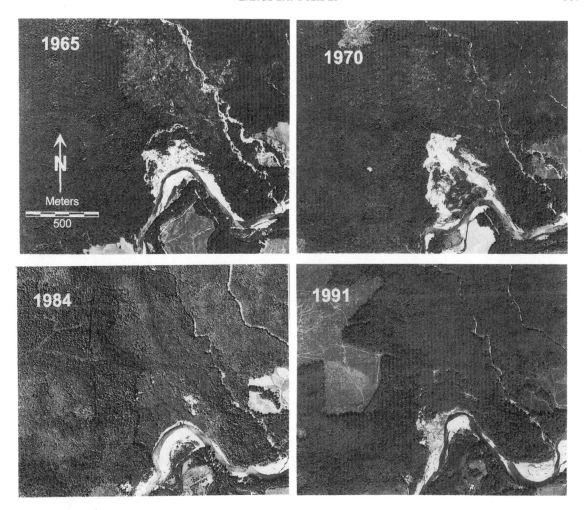

Figure 2. Aerial photography showing the landslide and vicinity. Large slump blocks and downslope flows are visible in the 1965 photograph. Also note the recent timber harvest of the small Headache Creek basin north of the landslide in this photo. A large event in 1967 dramatically altered the surface topography of the landslide and pushed the Stillaguamish River channel to the south, as seen in the 1970 photograph. The landslide scar revegetated over time, but several upslope areas exhibited persistent activity, seen as exposed soil in the 1984 photograph (the marks on the original have no bearing on this discussion). Harvest on the Whitman Bench and renewed slumping along the toe over the west end of the landslide are visible in the 1991 photograph

MODELS

With a GIS we create a spatially explicit representation of the topography and stratigraphy for the area surrounding the Hazel Landslide. Topography, illustrated in Figure 1 with shaded relief images, is represented by a digital elevation model (DEM) consisting of elevation values over a regular square grid of points (a 10 m grid spacing was used for this study) interpolated from 20-foot (6·1 m) contours on the 1:4800-scale topographic map. Stratigraphy is represented over the same grid with estimated elevations of the sand–clay contact, which is down-dropped in places by large slump blocks. Overlain on this landscape model are vegetation cover and precipitation values, both of which may vary in space and time. A hydrological model estimates groundwater recharge at each grid point as a function of vegetation cover, precipitation and temperature. The grid defines a finite-element mesh for a groundwater model, with recharge values over the grid specifying groundwater influx. Calculated water table elevations specify head values at each grid point, which are used in calculations of slope stability. Factors of safety are calculated along linear slope transects to estimate the minimum stability for every grid point. We then alter the vegetation cover to represent timber harvest (or alter the topography to represent bank erosion or channel incision) and repeat the exercise to find the relative decrease in stability at each point occurring in response to the applied change.

There is a vast range of hydrological, groundwater and slope stability models from which to choose. We are restricted by the data and resources at our disposal. Available data do not include geotechnical tests, water table elevations or extensive subsurface investigations, so we must keep the required number of model parameters to a minimum. This is an exploratory analysis intended, in part, to evaluate the necessity of further study, so low cost is another goal. We therefore selected inexpensive software to run on personal computers. The GIS we use is IDRISI (Clark Labs, 1997); the other models are described briefly below and in greater detail in Miller and Sias (1997).

Hydrology

Empirical paired catchment studies provide some indication of the magnitude of the change in recharge that occurs following timber harvest (Bosch and Hewlett, 1982; Stednick, 1996). These studies, however, show that site-specific factors are important determinants of the effects of timber harvest on water yield. A survey of rainfall and stream flow records in the region revealed no station pairs from which we might be able, at least roughly, to estimate the average annual evapotranspiration for a small forested and a small deforested catchment having zonal climate and geology similar to that of Hazel.

Faced with the task of estimating annual and, for transient analyses, shorter term recharge without the benefit of suitable paired catchment rainfall–runoff data, we chose to perform a hydrological simulation using a model adapted from Kelliher *et al.* (1986). These authors used the Penman–Monteith equation (Monteith, 1965) to estimate growing season evapotranspiration (ET) for a 31-year-old thinned Douglas fir stand on eastern Vancouver Island, British Columbia, Canada. This and numerous other studies (e.g. Rutter *et al.*, 1971; McNaughton and Black, 1973) have demonstrated that, for a wide variety of climates and vegetation covers, the Penman–Monteith equation can provide reasonably good estimates of growing season ET using *a priori*-determined soil and vegetation parameters. In particular, the study by Kelliher *et al.* (1986) shows that good results can be obtained for a Pacific Northwest coniferous forest.

The modified model uses a simplified representation of soil hydraulics (a so-called bucket model). The only required soil properties are the soil moisture content at field capacity and permanent wilting point. Soils (glacial sediments) at the Hazel site are deep and well drained, and the terrain is level over the majority of the groundwater recharge area; therefore, we did not include overland flow and lateral shallow subsurface flow in the model. Under these conditions, estimation of recharge as a function of vegetation cover reduces to a problem of estimation of evapotranspiration.

The model parameters are listed in Table I. Because we are interested in the effect of end-member vegetation cover on recharge, the model parameters are all vegetation dependent. As field determination of vegetation parameters was not feasible, we estimated from the literature what we consider to be reasonable upper and lower parameter bounds for a 'typical', non-species-specific coniferous forest (i.e. 20–40 m tall, single-layer, closed canopy) and a revegetated clear-cut (i.e. having a short, deciduous canopy and no overstorey) in the Pacific Northwest region. Model parameter values are given in Table I. For application to winter (October–March), leaf area index (LAI) was assumed to fall to nearly zero for the clear-cut and to decrease slightly for the forest. All other vegetation parameters are seasonally invariant.

Uncertainty as to the values to assign to the vegetation parameters in the evapotranspiration model translate to uncertainty in calculated recharge. For each cover, a high and low estimate for recharge was obtained by setting all parameters to their lower and upper bounds, respectively [parameters were defined such that upper (lower) bound values corresponded to high (low) evapotranspiration rates]. Climate was assumed invariant with changes in vegetation.

Table I. Model parameters, upper and lower parameter bounds and intermediate values

Parameter	Units	Evergreen forest			Clear-cut	
		Lower	Middle	Upper	Lower	Upper
r_a	s/cm	0·1	0·07	0·04	0·3	1·2
r_s-min	s/cm	2·5	2·5	2·5	2·0	1·5
LAI	m^2/m^2	5	7	10	0·75	1·50
FTHRU	Dimensionless	0·8	0·70	0·65	0·90	0·80
RZ-max	mm	200	250	300	100	150

r_a, aerodynamic resistance; r_s-min, minimum stomatal resistance; LAI, leaf area index; FTHRU, fraction throughfall plus stemflow; RZ-max, maximum root zone storage (storage at field capacity)

Evapotranspiration was calculated at a 6-hour time-step, as better results are obtained with the Penman–Monteith equation when the diurnal cycle is represented (McNaughton and Black, 1973). Six-hour meteorological variables (vapour pressure deficit, net radiation and air temperature) were inferred from daily precipitation and daily minimum and maximum temperature (T_{min} and T_{max}). Daily rainfall amounts were treated as 18-hour storms of constant intensity (on rainy days, the period 6 p.m. to midnight was taken to be rain free). We used a sinusoidal function to represent the diurnal variation in air temperature, the method of Running et al. (1987) to estimate vapour pressure deficit from T_{min} and T_{max}, and the algorithm of Bristow and Campbell (1984) to estimate daily insolation. Albedo was taken to be 0·12 and 0·18 for forest and clear-cut, respectively (Running et al., 1987). Long-wave radiation, based on air temperature and climatological averages of monthly cloud cover, was assumed to be constant over each 24-hour period, whereas insolation was assumed to be zero from 6 p.m. to 6 a.m. throughout the year. Because wind speed data were unavailable, aerodynamic resistance was treated as a vegetation parameter. Further details regarding the hydro-logical model (structure, parameter determination, elaboration of forcing data) can be found in Miller and Sias (1997).

Groundwater

We used MODFE, a groundwater model written by and available from the US Geological Survey. The program is freely available (Torak, 1993a; also over the internet at http://h2o.usgs.gov/software), well documented (Cooley, 1992; Torak, 1993b) and has a history of use (Czarnecki and Waddell, 1984; Buxton and Modica, 1992; Iverson and Reid, 1992; Reid and Iverson, 1992; Torak et al., 1992). MODFE is a two-dimensional finite-element model using linear, triangular elements. It can simulate both steady-state and transient groundwater flow in confined or unconfined aquifers and can accommodate seepage at the ground surface. We used the program to simulate unconfined aerial groundwater flux and to calculate water table elevations.

To represent aerial groundwater flux, the program assumes horizontal flow. The aquifer is characterized in terms of horizontal transmissivity, a measure of the potential horizontal flux of groundwater below a point on the ground surface, and specific yield, a measure of the change in volume of water stored within the aquifer caused by a unit change in water table elevation. Transmissivity is calculated as the depth integration of saturated hydraulic conductivity. In its original form, MODFE assumes a vertically homogeneous aquifer. For use at the Hazel Landslide site we modified the calculation of transmissivity to accommodate multiple layers.

The data required to describe the aquifer for this model are: (1) elevation of the ground surface; (2) elevation of stratigraphic contacts separating materials having differing hydraulic conductivity; and (3) hydraulic conductivity and specific yield for each of the materials. Ground surface elevations to construct a DEM are obtained from topographic mapping. Stratigraphic contacts and material properties of the aquifer had to be estimated. Stratigraphy was inferred from field-mapped contacts and surface morphology.

Surface groundwater seeps also proved useful for estimating contact locations. Hydraulic conductivity and specific yield values are constrained to some extent by the soil and rock types present. These values were adjusted during calibration runs of the model so that surface seepage is predicted in the same localities as observed. All data values are specified over a square grid of points within the GIS (IDRISI) and may vary spatially. We wrote pre- and post-processors using the IDRISI data structure so that MODFE was run directly from IDRISI files.

The grid points served as nodes of the finite-element mesh. A zero-flux boundary condition is enforced on the upslope boundary of the mesh, causing it to act as a water table divide. The upslope boundary was positioned along the drainage divide inferred from surface topography. In test runs of the program, the boundary was moved slightly to ensure that its location had no effect on calculated head values in the vicinity of the landslide. The remaining portions of the mesh boundary were positioned along a stream channel to the side of the landslide (Rollins Creek in Figure 1) and along the Stillaguamish River at its base, both inferred to be effluent from field observations of year-round seepage at their banks. Fixed-head boundary conditions were enforced along these portions of the mesh boundary.

Recharge values were specified for all interior nodes of the mesh. For a given pattern of recharge, the model calculates head elevations for all nodes. These elevations define a water table surface. Discharge values, to simulate seepage, were specified wherever the water table was at the ground surface (no recharge occurs at those nodes). A steady-state analysis produces a single water table surface corresponding to continuous recharge constant over time; a transient analysis produces a water table surface for each time step and illustrates aquifer response to changes in recharge over time. The head values are used by the stability model described below and to determine groundwater flow directions, from which the recharge area to specific points can be delineated.

The aerial extent of the recharge area must be inferred from model results. The calculated water table elevations were used with algorithms described in Jenson and Domingue (1988) for determining flow direction and delineating watersheds. The groundwater divide located for drainage to the landslide delineates the recharge area (Figure 3). We found that the spatial extent of the estimated recharge area varied little with temporal changes in recharge. We did not examine the effects of spatial variations in recharge.

Slope stability

Large landslides often exhibit a variety of mass-wasting processes. For example, the Hazel Landslide has experienced large slumps with rapid movement over a relatively planar slip surface. It also contains many rotational slumps of varying size, some of which fail catastrophically and some of which move incrementally year by year. Slump blocks may remain relatively coherent, with intact stratigraphy, others disintegrate into flows.

No single model can accurately portray all these types of behaviour, so, instead of a detailed landslide representation, we rely on a simple characterization of the potential for downslope mass movement. Gravity acts to move material downslope; movement is resisted by material shear strength. We estimate the relative magnitude of these forces by assuming limit equilibrium along distinct slip surfaces, using Bishop's simplified method of slices (Bishop, 1955) to calculate a factor of safety along individual slope transects. This procedure ignores the process of mass movement and focuses instead on the potential for movement.

We make no prior assumptions about the location of potential slip surfaces. For each slope transect, we use numerical minimization to find the set of slip surfaces giving the lowest factor of safety for each point along the transect (described in Miller, 1995). This procedure produces a 'factor of safety' profile, as shown in Figure 4. Each transect entails a multitude of potential slip surfaces; each slope entails a multitude of potential transects. We construct a linear transect from every grid point, oriented parallel to the slope aspect at that point and extending the length of the project area. Each grid cell (a cell is the surface area associated with a grid point) is thus intersected by numerous transects. Each point is assigned the lowest factor of safety found within its cell area. This procedure estimates the aerial pattern of relative slope stability.

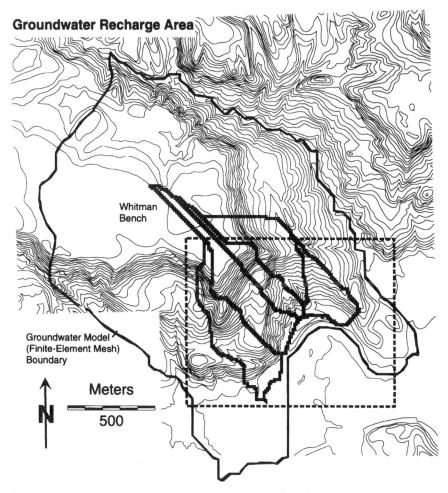

Figure 3. Estimated groundwater recharge area of the landslide. Heavy lines outline adjacent polygons that delineate five distinct zones of groundwater flow to the landslide, based on a steady-state simulation for fully forested conditions. The dashed rectangle outlines the area shown in Figures 4 and 6

The use of Bishop's simplified method of slices entails a circular slip surface and horizontal interslice forces. This method is numerically robust, which is why we use it, but it does not represent the least-stable geometry as well as other methods may (see, for example, Duncan and Wright, 1980). A circular shape still provides a great range of slip surface locations to examine, since the radius may vary to infinity. We do include a planar surface in the minimization scheme. A vertical tension crack is also allowed at the head scarp in every case. The depth of the tension crack is an adjustable variable in the search for the minimum factor of safety. We have experimented with other geometries. Spencer's method (Spencer, 1967) required a fourfold increase in processing time, and produced slightly lower factor of safety values in some areas. The relative change in the predicted factor of safety for different scenarios was, however, negligibly different from that obtained with Bishop's method.

We use a two-dimensional model, which implies plane strain (in the vertical plane). Extension of these methods to a three-dimensional analysis, although feasible (e.g. Hungr, 1987), is beyond our computing capacity. In general, a three-dimensional analysis produces larger factor of safety estimates than a two-dimensional analysis (see review, Duncan, 1992). We are unsure, however, of how three-dimensional effects would change the computed change in the factor of safety for different scenarios. In our current scheme, the

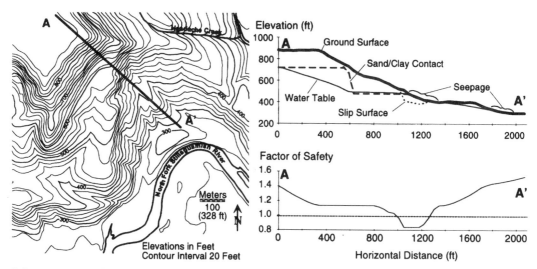

Figure 4. One transect used for calculating slope stability. Elevations of the ground surface, underlying stratigraphic contacts and water table are interpolated at equal intervals (10 m) from GIS grid files. These data are used to find the radius of the least-stable slip surface between every combination of interpolation points, from which the stability of the slope is estimated for each point along the transect, shown in the lower graph. The least stable slip surface found along this transect is shown in the upper graph. Thousands of such transects are used to estimate spatial patterns of slope stability

stability estimates for adjacent points along a slope contour are in no way connected; a three-dimensional analysis would include such interactions. We expect that a two-dimensional analysis, in some cases, underestimates the aerial extent of slope response and overestimates its magnitude.

The data required are: (1) ground surface elevations; (2) stratigraphic contact elevations; and (3) property values (cohesion, angle of internal friction and saturated and moist bulk densities) for each material. Determination of surface elevations and stratigraphic contacts was discussed in the description of the groundwater model above. Because these methods are intended for areas having little or no geotechnical data, material properties must be estimated. The rock and soil types present broadly limit the range of potential values. Specific values must be chosen through back-calculation over selected slope transects. We examined transects on both apparently stable and actively failing slopes that traversed all the stratigraphic units present (sand and clay). Using groundwater head values obtained for a typical wet season, we sought a combination of property values that produced factors of safety greater than 1 for stable slopes and less than 1 for failing slopes. The final values are reported in Table II.

We were not entirely successful in finding a single combination of property values appropriate for every slope examined. Material heterogeneity, particularly within a single stratigraphic unit, makes such success unlikely. The difference between peak and residual strengths can be large. [Palladino and Peck (1972) in examination of a deposit similar to that found at the Hazel site, report a peak strength represented by a cohesion of 0·65 ton/ft² (62 kPa) and a friction angle of 35° for intact clay, compared with a residual strength

Table II. Geotechnical properties

Property	Outwash sands	Lacustrine clays
Saturated conductivity (m/h)	0·36	0·00036
Specific yield	0·25	0·1
Bulk density (dry) (kg/m³)	1870	2020
Bulk density (saturated) (kg/m³)	2120	2120
Angle of internal friction (°)	38	23
Cohesion (kPa)	5	14

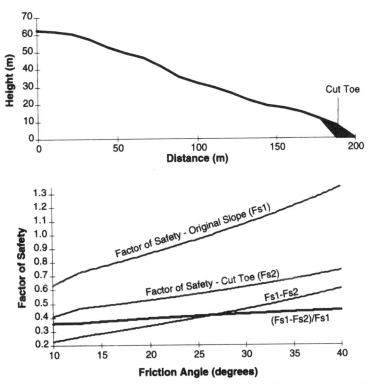

Figure 5. Proportional response as a measure of landslide sensitivity that is relatively insensitive to errors in estimated geotechnical properties

represented by a cohesion of zero and a friction angle of from 13·5 to 17·5° for disturbed clay.] The transects used for back-calculating property values traverse slopes of both disturbed and intact material. Over disturbed slopes, the depth to intact material is unknown. The values we chose are thus a compromise between peak and residual strengths, so that factors of safety for intact slopes are systematically under-estimated and those for failed slopes are systematically overestimated.

Unresolved spatial heterogeneity is a confounding factor that renders predicted factor of safety values uncertain. Unfortunately, we see no means of mapping heterogeneity inexpensively (in three dimensions) at the scale of individual slump blocks. We therefore minimize reliance on factor of safety estimates by examining the change, rather than the actual values themselves, predicted for some change in a model variable. The proportional decrease in stability caused by an increase in recharge or cutting of the slope toe, for example, is relatively independent of the geotechnical values assigned to the material. This is illustrated in Figure 5, which shows the factor of safety calculated for a slope entirely in clay. Stability of the slope both before and after cutting of the toe, and the difference between these two cases, is highly dependent on the friction angle assigned to the clay. The proportional change, however, varies over a much smaller range. Hence we use the proportional change, calculated as

$$\frac{(Fs_1 - Fs_2)}{Fs_1} \qquad (1)$$

(where Fs_1 and Fs_2 are the factors of safety before and after a perturbation, respectively), as an indicator of slope response that is less sensitive to errors in the assigned geotechnical properties than the factor of safety values themselves. We use Equation (1) to evaluate and compare the effects of changes in slope topography (cutting of slope toes) and changes in pore pressure (associated with changes in groundwater recharge).

Variation of pore pressures is a primary control on slope stability. We are interested in both spatial and temporal variability. Pore pressures in the model are defined by the head values obtained with the groundwater model. In using a two-dimensional groundwater model that approximates only the horizontal components of the flow field, we focus on the aerial variation in hydrostatic pressure, but miss the effect on slope stability of head gradients in the vertical plane, which are substantial on slopes and across contrasts in hydraulic conductivity (Iverson and Reid, 1992; Reid and Iverson, 1992). When examining slope response to relatively small head changes, with no change in topography, the effects of this shortcoming should be minimized by use of Equation (1). When the topography is altered by cutting of the slope toe, consequent changes in vertical components of the flow field, which are not represented here, may have a larger effect on slope response, primarily near the toe. We have not yet quantified these effects within the context of the models used here. We expect that estimates of slope response to both pore pressure changes and to alterations of slope topography are, in some areas, underestimated.

RESULTS: STEADY-STATE AND TRANSIENT ANALYSES

Groundwater recharge

For a given climatic time-series, the hydrology model provides a time varying, point estimate of recharge to groundwater as a function of vegetation cover. Although vegetation may vary both spatially and temporally within the model, we chose to examine two scenarios having vegetation attributes constant in both space and time: fully forested and fully clear-cut. More realistic harvest scenarios that include limited patches of clear-cutting and forest regrowth could be simulated; our goal here was to estimate the maximum effect of clear-cutting over any time interval throughout the climatic time-series.

The model predicts that average annual forest evapotranspiration at Hazel lies between 45 and 75% of annual rainfall, and that winter evapotranspiration accounts for 50% or more of annual evapotranspiration. This result is a consequence of above freezing winter temperatures at the Hazel site. Lower temperatures would result in less winter evapotranspiration. The predicted annual interception loss for forest ranges from 312 to 552 mm. [This prediction compares well with the results of a recent US Geological Survey study, in which annual interception loss for a Puget Sound lowland forested plot, situated on glacial till, was found to be 450 mm. Annual rainfall was about 1000 mm at that site (Henry Bauer, personal communication).] Annual evapotranspiration for the clear-cut amounts to about 20% of annual rainfall. Summer evapotranspiration is quite similar for both forest and clear-cut. Winter evapotranspiration for the clear-cut is negligible.

The difference between the forest high recharge case (923 mm/yr mean value) and the clear-cut low recharge case (1204 mm/yr) forms the low estimate for change in recharge due to clear-cutting (281 mm/yr). The difference between the forest low recharge case (512 mm/yr) and the clear-cut high recharge case (1375 mm/yr) forms the high estimate for change in recharge due to clear-cutting (883 mm/yr). The predicted change in average annual evapotranspiration owing to clear-cutting is mostly attributable to a nearly 100% decline in evapotranspiration in winter.

Landslide response to harvest-related increases in recharge

We used the time-averaged recharge values for the fully forested and fully clear cut scenarios with the groundwater model to estimate steady-state head values for each case. The estimated recharge area to the landslide varied negligibly between the two. The change in landslide stability was gauged for each point over the landslide using Equation (1). Because of uncertainty in the vegetation parameter values, this exercise was performed for both the smallest and largest estimated changes in recharge. The results are shown in Figures 6A and 6B.

High sensitivity to increases in recharge does not necessarily indicate that the landslide will be affected. A stable, but sensitive, slope will be stable in either case (and vice versa). Thus, grid points predicted to be

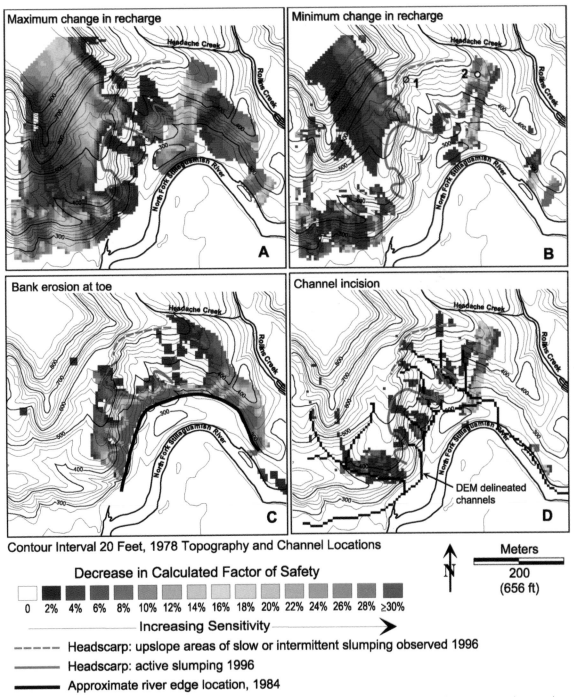

Figure 6. Changes of slope stability in response to (A) clear-cut timber harvesting of the groundwater recharge area, maximum estimate, (B) clear-cut timber harvesting of the groundwater recharge area, minimum estimate, (C) bank erosion of the landslide toe, and (D) incision of channels draining the body of the landslide. Zones of landslide activity observed during field visits in 1995 and 1996 are indicated by heavy brown lines. Points 1 and 2 shown in (B) are the locations of the transient stability analyses. Topography and channel locations are based on 1978 aerial photography

stable, as defined by a factor of safety of 1·3 or greater in the least stable case [Fs_2 in Equation (1)], are excluded from these maps.

The resulting maps of sensitivity indicate that certain portions of the landslide respond to changes in recharge differently than do other portions: many areas are unaffected, while others experience up to a 30% decrease in stability. The pattern revealed is a consequence of: (1) the spatial differences of the change in head values over the landslide between forested and clear-cut conditions; and (2) the spatial variability of surface and subsurface geometry, which produces a unique response from each slope to a change in head.

Despite our use of spatially invariant vegetation attributes for these analyses, one may use these results to estimate spatial correlations between areas of recharge and landslide response. Compare the pattern of sensitivity shown in the maps of Figure 6 with the recharge zones delineated for various portions of the landslide in Figure 3. Different parts of the landslide are fed by separate recharge zones; hence, a change in vegetation through a portion of the recharge area, a harvest unit for example, should primarily affect only a certain portion of the landslide.

Response to cutting of the slope toe

To evaluate the effects of bank erosion at the toe, we simply 'removed' material from the model and moved the northern river edge to its approximate 1984 location. We then recalculated the factors of safety for this modified topography and used Equation (1) as a measure of the change. We used forested recharge values (high estimate) for each case. The pattern of response shown in Figure 6C is a function of the surface and subsurface geometry represented in the model.

Increased recharge and bank erosion at the toe both cause spatially variable effects, with each primarily affecting different portions of the landslide. Near the toe, bank erosion lowered the estimated stability by nearly 75%, a much larger effect than that associated with increased recharge. Upslope, the largest change is associated with increasing recharge, although the maximum indicated reduction is 30%, less than half of the maximum estimated for toe cutting.

Response to incision of channels draining the landslide

Several small streams drain the body of the landslide. Numerous adjacent slumps and persistent mud flows feed into their channels. Evidence of channel incision (noted also by Shannon and Associates, 1952) and bank-cutting suggests that mass wasting is initiated and frequently reactivated by fluvial erosion of the channels. Small-scale slumping into these channels occurs throughout the winter months, with fluvial flushing of the fine-grained debris into the Stillaguamish occurring year-round. Over time these minor topographic readjustments and the continual removal of material may act to destabilize larger portions of the landslide, thus activating larger slumps. We evaluate the effects of this process on landslide stability below.

We use the 10-m grid DEM to delineate channels on the landslide as shown in Figure 6D. Channel head locations are estimated from their current field locations. We then digitally incise every channel by one meter and recalculate factors of safety over the landslide. The effects of this incision are displayed, via Equation (1), in Figure 6D.

Discussion of steady-state analyses

Aerial photographs and previous reports (Figure 2; Shannon and Associates, 1952; Benda *et al.*, 1988) document landslide activity over the last 50 years and are summarized in Figure 7A. Topographic data for this study were based on 1978 aerial photography, so some interpretation is required in applying our results to the landslide over time. A large 1967 event dramatically altered the surface geometry of the landslide and calculated stability and sensitivity will not apply before that time. Topographic changes since 1978, such as northward migration of the river channel and slumping along the western portion of the landslide, may also alter landslide responses from those suggested in Figure 6.

Nevertheless, observed landslide activity correlates well with areas predicted to respond to one or more of the applied changes. Upslope areas of activity seen in post-1970 photographs and in the field (1995–1996)

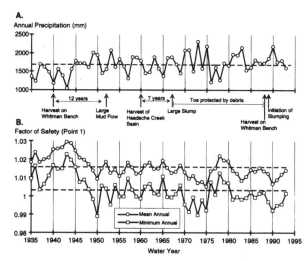

Figure 7. (A) Annual precipitation over the period of record. Timing of timber harvest and landslide activity are also indicated. (B) Estimated mean annual and minimum annual factor of safety over the period of record for point 1 in Figure 6B. Estimated slope stability varies inversely with annual precipitation, but also shows longer term variation. Stability at any time depends on the history of recharge over multiple years

fall within areas indicated to be sensitive to increases in recharge and to incision of channels draining the body of the landslide (compare light patches, indicating areas of exposed soil in the photographs, in Figure 2 with the maps in Figure 6). Large riverside slumps in 1988, and again since 1996, occurred in areas sensitive to cutting of the toe. Extensive slumping in an area sensitive to channel incision occurred during a large winter storm in 1996. The predicted pattern and relative magnitude of landslide response to these different environmental factors offer hypotheses to explain observed landslide activity.

However, many of the sensitive areas shown in Figure 6 show no signs of instability. Slopes up to the Whitman Bench, for example (the large red to yellow area in the north-western portion of Figure 6A), appear, on examination in the field, to be quite stable. The predicted high sensitivity of stable slopes is a consequence of two factors: a stable slope may experience a decrease in the factor of safety and still be stable and, because material strength parameters are underestimated for undisturbed slopes, the stability for these slopes is also underestimated. Predictions of sensitivity to change in a particular model variable can serve to elucidate field observations, but must be interpreted within the context of those observations.

Effects of toe cutting. Landslide history clearly indicates great dependence on conditions at the base along the Stillaguamish River. Big, river-moving events have occurred only when the channel impinges on the toe. No large events took place during the rather wet years of the early and mid 1970s when the toe was protected by debris from the 1967 event (Figure 7A). It was not until 1988, when that debris was eroded away and the river again had access to the toe of the landslide, that large-scale, riverside slumping reoccurred. At the base of the slope, the calculated effects of toe erosion are many times greater than those of increased recharge or channel incision, and the zones of observed activity match well with those predicted to be sensitive to toe cutting.

Even during times of apparent quiescence, however, activity persists upslope, as shown by the persistent appearance of exposed headscarps and unvegetated areas in 1978, 1984 and 1988 photographs. Several areas of upslope activity were observed during field visits in 1995 and 1996 (shown in Figure 6). In general, upslope areas of observed landslide activity correlate well with areas predicted to be sensitive either to increases in recharge and/or to incision of channels draining the landslide.

Effects of timber harvest. Benda *et al.* (1988) noted: (1) the increase of landslide activity in the early 1950s following timber harvest on Whitman Bench in 1940; and (2) the increase of landslide activity in the mid-1960s following harvest of the Headache Creek basin in 1960. We now add, (3) slumping in the western portion of the landslide shown in the 1991 photograph following timber harvest within the groundwater recharge area on Whitman Bench in 1988.

We can compare the location of the harvest to the subsequent locus of activity on the landslide. Based on the recharge areas delineated for different portions of the landslide in Figure 3, the harvest of Headache Creek basin, as occurred around 1960, should primarily affect the east-central portion of the landslide, correlating with the location of the large 1967 event (compare the landslide scar in the 1970 photograph of Figure 2 with the polygons shown in Figure 3). Harvest of the Whitman Bench, as occurred in the late 1980s, should primarily affect slopes through the west and west-central portions of the landslide, correlating with activity in the 1990s. Model predictions suggest that logging activity in certain portions of the recharge area will affect certain portions of the landslide. Such a cause and effect relationship is not refuted by the observed spatial correlations between logging and landslide activity.

One must examine these correlations in relationship to the pattern of annual precipitation (Figure 7A). The first reported period of landslide activity (1950s) started during a year of below average rainfall. The other two periods started during years of slightly above average rainfall. The first two periods of landslide activity began 12 and 7 years after harvest, the third about a year after harvest. Interpretation of these observations requires a look at the temporal nature of landslide response.

Temporal patterns of slope response

Groundwater flux to the landslide varies over time in response to changing rates of recharge. Recharge varies both seasonally, being greater in winter than in summer, and annually, being greater in wet years than in drought years. The variability in recharge caused by annual variability in precipitation is larger than the change estimated for conversion from forested to clear-cut conditions. Any change in recharge caused by logging must be assessed in conjunction with natural year to year variability in precipitation.

The time variability of groundwater flux and consequent changes in slope stability can be simulated with the models presented here. Unfortunately, our computing capacity was insufficient to generate a factor of safety time series for all points on the landslide. It is, nevertheless, instructive to examine temporal changes at a limited number of points.

Transient groundwater simulations indicated that the aquifer responds to multiyear patterns of precipitation. A series of wet years can elevate the water table for several subsequent dry years, and vice versa. For the points examined, the calculated factor of safety varied inversely with the volume of water in aquifer storage, which varies directly with multiyear patterns of precipitation. The predicted time-series of stability, drawn for a single point in Figure 7B, shows annual variability driven by year to year changes in rainfall, overlain on longer term variability driven by multiyear patterns of rainfall.

Regression of the factor of safety to total annual precipitation for previous years indicates that slope stability at this point responds to the previous five years of precipitation (Figure 8). The length of this memory effect is a function of the size and shape of the recharge area and of aquifer geometry and hydraulic conductivity. Our estimates for all of these values are of uncertain accuracy, so five years may be either an under- or over-estimate. Nevertheless, this result illustrates a potentially important aspect of landslide behaviour. Slopes coupled to regional groundwater flow can respond to multiyear patterns of recharge.

Although we were only able to estimate the length of landslide 'memory' for one point on the landslide, we expect that temporal patterns of landslide response vary spatially. We examined the stability of two points over the course of a single water year, the results of which are shown in Figure 9. The graphs show cumulative factor of safety frequency distributions at each point for both forested and clear-cut cases. Both points lay near the heads of small slump blocks identified in the field. Both slumps have headscarps within the sandy outwash and extend into the lacustrine deposits. Both are associated with groundwater seeps. Greatly disrupted stratigraphy, trees growing at a variety of angles and a complete lack of conifers indicate prolonged slow movement of slump block 1. Less disruption of stratigraphy and the presence of conifers suggests less movement of slump block 2, although tree growth angles still indicate a long history of movement. Despite

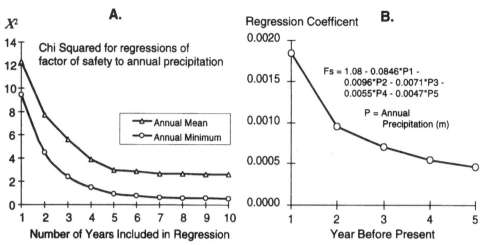

Figure 8. Stability dependence on multiyear precipitation. (A) Multiple regression of the factor of safety (at point 1 in Figure 6B) with annual precipitation. The error decreases as additional years are included in the regression, up to about five years, after which the error levels off. (B) Coefficients for the five-year regression. The current year has the greatest individual effect, but the previous four years have an overall greater influence

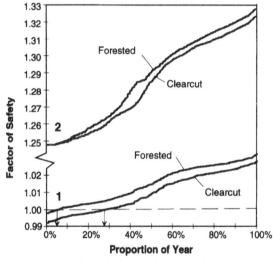

Figure 9. Factor of safety frequency distributions for points 1 and 2 (Figure 6B) over water year 1989. Results for both forested and clear-cut cases are shown. A small decrease in stability at point 1 increases the proportion of the year spent in a state of failure from 5% to 28%. Differences in the factor of safety values are of less importance than differences in the frequency distribution shape. Stability at point 2 varies over a greater range and responds differently to changes in recharge than that at point 1

obtaining factor of safety values always greater than unity for point 2, we surmise that both points dip below unity some years. Because of the problems with unresolved spatial variability in material properties discussed earlier, the difference in factor of safety values between the two points is of less importance to our analysis than the difference in their temporal patterns of change.

Failure in these cases entails a balance of forces that allows slow or intermittent mass movement. We surmise that the extent of movement depends not only on the frequency with which the factor of safety dips below unity, but also on the amount of time that it stays there. Any perturbation that lowers the stability of a marginally stable slope increases the proportion of time spent in a state of failure. As shown for point 1, a decrease in stability of only half a per cent increased the proportion of the year spent in a condition of slope mass movement from 5 to 28%.

This result depends strongly on the shape of the frequency distribution. If the graph for point 1 were steeper near the point of instability, the increase would be less. Differences in predicted frequency distribution shapes for points 1 and 2 indicate that the temporal response of the landslide may vary from point to point in a manner unrelated to the magnitude of the change. The nature of the response to changing recharge also differs between the two points. The minimum at point 2 is unchanged between forested and clear-cut conditions. (This occurs because the slump block containing point 2 becomes completely saturated in both cases.) To characterize landslide behaviour in terms of mass-wasting sediment flux, we must examine both spatial and temporal patterns of slope response to changes in groundwater recharge (and/or other factors). The predicted magnitude of the response shown for the steady-state analyses in Figure 6 tell only part of the story. (We also note that the transient analysis suggests a smaller magnitude of stability reduction than indicated by the steady-state results.) These results suggest that it would be instructive to map some integrated measure of temporal response over the landslide.

Discussion of transient analyses

The various time lags observed between timber harvest in the groundwater recharge area and renewed activity on the landslide can be examined in light of the transient analyses discussed above. The dependence of predicted stability on multiyear patterns of rainfall reflect the time taken for water to travel through the aquifer. Transit times are a function of head gradients, which change over time as the aquifer drains through surface seeps and is recharged from above. A localized increase in recharge associated with timber harvest in a portion of the recharge zone will locally elevate the water table. The associated change in groundwater flow is communicated through the aquifer at a rate depending on the magnitude of the water table rise and on the hydraulic conductivity of the material. The results shown in Figure 8 suggest that it may take some years for head values to respond to changes across the aquifer, so that a lag between timber harvest and landslide response may be expected. We suspect that the response time varies spatially and temporally, depending on the spatial and temporal patterns of recharge. This study did not include a simulation of harvest history imposed on the time-series of recharge. Such an analysis may prove useful for estimating the relative response times for each case and, thereby, for better evaluating the degree to which past timber harvests were related to subsequent landslide activity. (Such an analysis would also provide a temporal test of the models, since we could compare the sequence of predicted landslide events to those observed.) All we can say at this time is that we expect some lag time between timber harvest-related increases in recharge and landslide response, and that the length of that lag may vary between cases, depending on the location of the timber harvest and on the sequence of precipitation events.

CONCLUSIONS

This study was undertaken to aid regulators and land managers in assessing the risk posed by timber harvest within the groundwater recharge area of a large landslide complex. Using GIS to link simple hydrological, groundwater and slope stability models using available data, we estimated the spatial patterns and relative magnitudes of potential landslide response to changes in three specific factors: (1) altered groundwater flow associated with timber harvest of the recharge area of the landslide; (2) bank erosion of the landslide toe; and (3) incision of channels draining the body of the landslide. Given the limitations of the models used and uncertainties in the material parameters chosen, we found it useful to examine the relative change in stability predicted for each case, rather than the factor of safety values obtained. These results were based on steady-state analyses and matched well with observed locations of landslide activity.

A transient analysis of slope stability performed for two points on the landslide illustrate temporal patterns of landslide behaviour (limited computer resources prevented a more extensive transient analysis). We find that stability varies inversely with annual precipitation, with additional dependence on precipitation patterns extending over several years (five in the example here). We show how a small reduction in the stability of marginally stable slopes may greatly extend the length of time they are in a state of failure, which, for slow-moving slumps, may cause an increase of mass flux much larger than indicated by the proportional change in stability. The increased time spent in motion (failure) caused by a given reduction in stability may vary, however, from point to point over the landslide.

These results offer hypotheses to explain observed patterns of landslide behaviour. As such, they provided decision makers with improved understanding of the factors affecting slope stability and of the potential relationships between different processes active at the Hazel Landslide. Because of results of this study, future timber harvest plans at the Hazel site must be designed (through, for example, partial cuts, phased clear-cuts) to have no anticipated effect on landslide stability, and must include a detailed analysis showing how the harvest plan avoids such effects. Ensuing discussions led to acknowledgment of the uncertainties involved and of the potential to address these issues better with additional information. Hence, landslide, precipitation, and groundwater monitoring must also accompany any future timber harvests approved for this site.

ACKNOWLEDGEMENTS

Thanks to all involved with the Hazel Watershed Analysis team for their willingness to try a new approach to an often contentious land use issue. Paul Kennard, geologist with the Tulalip Tribes, and Bob Penhale, at the Washington Department of Ecology, in particular had the vision and tenacity to rally support for this undertaking. This paper benefitted greatly from reviews by D. R. Montgomery, K. M. Schmidt and one anonymous reviewer. Funding was provided by the Washington Department of Ecology and administered by the Tulalip Tribes.

REFERENCES

Benda, L., Thorsen, G. W., and Bernath, S. 1988. 'Report of the I.D. team investigation of the Hazel Landslide on the North Fork of the Stillaguamish River', *Report F.P.A. 19-09420*. Report to the Washington Department of Natural Resources, Northwest Region, Sedro Woolley, Washington. 13 pp.

Bishop, A. W. 1955. 'The use of the slip circle in the stability analysis of slopes', *Geotechnique*, **5**, 7–17.

Booth, D. B. 1989. *Surficial Geologic Map of the Granite Falls 15-Minute Quadrangle, Snohomish County, Washington*. US Geological Survey Miscellaneous Investigations Series, Map I-1852. USGS, United States Geological Survey, Washington, D.C.

Bosch, J. M. and Hewlett, J. D. 1982. 'A review of catchment experiments to determine the effect of vegetation changes on water yield and evapotranspiration', *J. Hydrol.*, **55**, 3–23.

Bristow, K. L. and Campbell, G. S. 1984. 'On the relationship between incoming solar radiation and daily maximum and minimum temperature', *Agric. For. Meteorol.*, **31**, 159–166.

Buxton, H. T. and Modica, E. 1992. 'Patterns and rates of ground-water flow on Long Island, New York', *Ground Water*, **30**, 857–866.

Clark Labs, 1997. *Idrisi for Windows*, Version 2·0. Clark Labs for Cartographic Technology and Geographic Analysis, Clark University, Worcester, Massachusetts.

Cooley, R. L. 1992. *A MODular Finite-Element model (MODFE) for areal and axisymmetric ground-water-flow problems, part 2 — derivation of finite-element equations and comparisons with analytical solutions*, US Geological Survey Techniques of Water Resources Investigations, Book 6, Chap. A4. USGS, Washington, D.C.

Czarnecki, J. B. and Waddell, R. K. 1984. 'Finite-element simulation of ground-water flow in the vicinity of Yucca Mountain, Nevada-California', *US Geological Survey Water Resources Investigations Report 84-4349*. USGS, Washington, D.C. 38 pp.

Dietrich, W. E., Reiss, R., Hsu, M. L., and Montgomery, D. R. 1995. 'A process-based model for colluvial soil depth and shallow landsliding using digital elevation data', *Hydrol. Process.*, **9**, 383–400.

Duncan, J. M. 1992. 'State-of-the-art: static stability and deformation analysis', in Seed, R. B. and Boulanger, R. W. (eds), *Stability and Performance of Slopes and Embankments II*, Geotechnical Special Publication 31. American Society of Civil Engineers, New York. pp. 222–266.

Duncan, J. M. and Wright, S. G. 1980. 'The accuracy of equilibrium methods of slope stability analysis', *Engng Geol.*, **16**, 5–17.

Hammond, C., Hall, D., Miller, S., and Swetik, P. 1992. 'Level 1 stability analysis (LISA) documentation for version 2.0', *General Technical Report INT-285*, US Department of Agriculture, Forest Service, Intermountain Research Station, Moscow, Idaho. p. 190.

Hungr, O. 1987. 'An extension of Bishop's simplified method of slope stability analysis to three dimensions', *Geotechnique*, **37**, 113–117.

Iverson, R. M. and Reid, M. E. 1992. 'Gravity-driven groundwater flow and slope failure potential, 1. Elastic effective-stress model', *Wat. Resour. Res.*, **28**, 925–938.

Jenson, S. K. and Domingue, J. O. 1988. 'Extracting topographic structure from digital elevation data for geographic information system analysis', *Photogramm. Engng Remote Sens.*, **54**, 1593–1600.

Kelliher, F. M., Black, T. A., and Price, D. T. 1986. 'Estimating the effects of understory removal from a Douglas fir forest using a two-layer canopy evapotranspiration model', *Wat. Resourc. Res.*, **22**, 1891–1899.

McNaughton, K. G. and Black, T. A. 1973. 'A study of evapotranspiration from a Douglas Fir forest using the energy balance approach', *Wat. Resour. Res.*, **9**, 1579–1590.

Miller, D. J. 1995. 'Coupling GIS with physical models to assess deep-seated landslide hazards', *Environ. Engng Geosc.*, **1**, 263–276.

Miller, D. J. and Sias, J. 1997. *Environmental Factors Affecting the Hazel Landslide*, Level 2 Watershed Analysis Report. Washington Department of Natural Resources, Northwest Region, Sedro Woolley, Washington.

Monteith, J. L. 1965. 'Evaporation and environment', *Symp. Soc. Exp. Biol.*, **19**, 205–234.

Palladino, D. J. and Peck, R. B. 1972. 'Slope failures in an overconsolidated clay, Seattle, Washington', *Geotechnique*, **22**, 563–595.

Pearce, A. J., Gash, J. H. C., and Stewart, J. B. 1980. 'Rainfall interception in a forest stand estimated from grassland meteorological data', *J. Hydrol.*, **46**, 147–163.

Reid, M. E. and Iverson, R. M. 1992. 'Gravity-driven groundwater flow and slope failure potential, 2. Effects of slope morphology, material properties, and hydraulic heterogeneity', *Wat. Resour. Res.*, **28**, 939–950.

Running, S. W., Nemani, R. R., and Hungerford, R. C. 1987. 'Extrapolation of synoptic meteorological data in mountainous terrain, and its use for simulating forest evapotranspiration and photosynthesis', *Can. J. For. Res.*, **17**, 472–483.

Rutter, A. J., Kershaw, K. A., Robins, P. C., and Morton, A. J. 1971. 'A predictive model of rainfall interception in forests, 1. Derivation of the model from observations in a plantation of Corsican Pine', *Agric. Meteorol.*, **9**, 367–384.

Shannon, W. D. and Associates, 1952. *Report on Slide on North Fork Stillaguamish River near Hazel, Washington, unpublished report to the State of Washington Departments of Game and Fisheries.* 18 pp.

Sidle, R.C., Pearce, A. J., and O'Loughlin, C. L. 1985. *Hillslope Stability and Land Use*, American Geophysical Union, Washington, DC. 140 pp.

Spencer, E. 1967. 'A method of analysis of the stability of embankments assuming parallel interslice forces', *Geotechnique*, **17**, 11–26.

Stednick, J. D. 1996. 'Monitoring the effects of timber harvest on annual water yield', *J. Hydrol.*, **179**, 79–95.

Swanston, D. N., Lienkaemper, G. W., Mersereau, R. C., and Levno, A. B. 1988. 'Timber harvest and progressive deformation of slopes in southwestern Oregon', *Bull. Assoc. Engng Geol.*, **25**, 371–381.

Tabor, R. W., Booth, D. B., Vance, D. A., and Ford, A. B. 1988. *Geologic Map of the Sauk River 30- by 60-Minute Quadrangle, Washington.* US Geological Survey Open File Report 88-692. United States Geological Survey, Washington, D.C. 50 pp., 1 map sheet.

Thorsen, G. W. 1989. 'Landslide provinces in Washington', in Galster, R. W. (ed.), *Engineering Geology in Washington*, Volume 1, Bulletin 78. Washington Division of Geology and Earth Resources, Department of Natural Resources. Olympia, Washington. pp. 71–89.

Torak, L. J. 1993a. *A MODular Finite-Element model (MODFE) for areal and axisymmetric ground-water-flow problems, part 1 — model description and user's manual*, US Geological Survey Techniques of Water Resources Investigations, Book 6, Chap. A3. USGS, United States Geological Survey, Washington, D.C.

Torak, L. J. 1993b. *A MODular Finite-Element model (MODFE) for areal and axisymmetric ground-water-flow problems, part 3 — design philosophy and programming details*, US Geological Survey Techniques of Water Resources Investigations, Book 6, Chap. A5. USGS, United States Geological Survey, Washington, D.C.

Torak, L. J., Davis, G. S., Herndon, J. G., and Strain, G. A. 1992. 'Geohydrology and evaluation of water-resource potential of the upper Floridan aquifer in the Albany area, southwestern Georgia', *US Geological Survey Water-Supply Paper 2391*. USGS, United States Geological Survey, Washington, D.C. 59 pp.

Varnes, D. J. 1978. 'Slope movement types and processes', in *Landslides — Analysis and Control*, National Academy of Sciences Transportation Research Board Special Report 176, Chap. 2. National Research Council, Washington, DC.

WDNR (Washington Department of Natural Resources), 1996. *Level 1 Hazel Watershed Analysis, Washington State*, Watershed Analysis Report, Washington Department of Natural Resources, Northwest Region, Sedro Woolley.

WFPB (Washington Forest Practices Board), 1995. *Standard Methodology for Conducting Watershed Analysis*, Version 3.0. Washington Forest Practices Board, Olympia, Washington.

Wolff, N., Nystrom, M., and Bernath, S. 1989. *The effects of partial forest-stand removal on the availability of water for groundwater recharge*, unpublished report, Washington Department of Natural Resources, Northwest Region, Sedro Woolley. 14 pp.

9

REGIONAL TEST OF A MODEL FOR SHALLOW LANDSLIDING

DAVID R. MONTGOMERY,[1*] KATHLEEN SULLIVAN[2] AND M. GREENBERG[1]

[1]*Department of Geological Sciences, University of Washington, Seattle, WA, USA*
[2]*Environmental Research, Weyerhaeuser Co., Tacoma, WA, USA*

ABSTRACT

Landslides mapped in 14 watershed analyses in Oregon and Washington provide a regional test of a model for shallow landsliding. A total of 3224 landslides were mapped in watersheds covering 2993 km^2 and underlain by a variety of lithologies, including Tertiary sedimentary rocks of the Coast Ranges, volcanic rocks of the Cascade Range and Quaternary glacial sediments in the Puget Lowlands. GIS (geographical information system) techniques were used to register each mapped landslide to critical rainfall values predicted from a theoretical model for the topographic control on shallow landsliding using 30 m DEMs (digital elevation models). A single set of parameter values appropriate for simulating slide hazards after forest clearing was used for all watersheds to assess the regional influence of topographic controls on shallow landsliding. Model performance varied widely between watersheds, with the best performance generally in steep watersheds underlain by shallow bedrock and the worst performance in generally low gradient watersheds underlain by thick glacial deposits. Landslide frequency (slides/km^2) varied between physiographic provinces but yielded consistent patterns of higher slide frequency in areas with lower critical rainfall values. Simulations with variable effective cohesion predicted that high root strength effectively limits shallow landsliding to topographic hollows with deep soils and locations that experience excess pore pressures, but that low root strength leads to higher probabilities of failure across a greater proportion of the landscape.

KEY WORDS   landslides; GIS; model testing, forest management

INTRODUCTION

Shallow landsliding can dominate sediment transport in steep, soil-mantled landscapes (Swanson *et al.*, 1982; Dietrich *et al.*, 1986; Benda, 1990; Seidl and Dietrich, 1992) and topographically-driven convergence of both soil and runoff favour the initiation of shallow landslides in fine-scale topographic hollows (Williams and Guy, 1971; Dietrich *et al.*, 1986). The downstream disturbance, scour and deposition from shallow landslides that transform into debris flows can adversely affect channels, people, fish and property. Land use can affect shallow landsliding and, even though many studies have demonstrated acceleration of landsliding following road construction and timber harvest (e.g. Fredriksen, 1970; Brown and Krygier, 1971; Mersereau and Dyrness, 1972; Swanson and Dyrness, 1975; Swanston and Swanson, 1976; Gresswell *et al.*, 1979), doubts are still voiced over the role of forest clearing on the initiation of shallow landslides (e.g. Skaugest *et al.*, 1993; Martin *et al.*, 1996). Historically, lack of methods to stratify equivalent topographic influences on shallow landsliding compromised evaluation of the effects of land management. GIS-driven models can provide a spatially distributed prediction of the relative role of topographic influences on shallow landslide

---

* Correspondence to: David R. Montgomery, Department of Geological Sciences, University of Washington, Seattle, WA 98195, USA.

Contract grant sponsors: Weyerhaeuser Company; Cooperative State Research Services; USDA; USFS; NFS.
Contract grant numbers: 94-37101-0321 (USDA); PNW 93-0441 (USFS); CMS-9610269 (NSF).

initiation (e.g. Dietrich *et al.*, 1993; Montgomery and Dietrich, 1994; Wu and Sidle, 1995), and an extensive programme of landslide mapping conducted during recent watershed analyses (WFPB, 1993) provides an opportunity for regional tests of such models. Here we use data from 14 watersheds in Oregon and Washington to examine the performance of a model for the topographic control of shallow landsliding.

MODEL

Landslide hazard assessments are based on a variety of approaches and assumptions. Many approaches rely on either multivariate correlations between mapped landslides and landscape attributes (e.g. Neuland, 1976, 1980; Carrara *et al.*, 1977, 1991; Carrara, 1983; Mark, 1992; Busoni *et al.*, 1995), or general associations of landslide hazard from rankings based on slope, lithology, land form or geological structure (e.g. Brabb *et al.*, 1972; Campbell, 1975; Hollingsworth and Kovacs, 1981; Lanyon and Hall, 1983; Seely and West, 1990; Montgomery *et al.*, 1991; Niemann and Howes, 1991; Derbyshire *et al.*, 1995). The approach explored here builds on the physics-based modelling pioneered by Okimura and colleagues (Okimura and Ichikawa, 1985; Okimura and Nakagawa, 1988) and extended by others (e.g. Dietrich *et al.*, 1993, 1995; van Asch *et al.*, 1993; Montgomery and Dietrich, 1994) to develop a simple model of pre- and post-cutting hazard from shallow landsliding.

Our approach is based on coupling a hydrological model to a limit-equilibrium slope stability model to calculate the critical steady-state rainfall necessary to trigger slope instability at any point in a landscape. The hydrological model maps the spatial pattern of equilibrium soil saturation based on analysis of upslope contributing areas, soil transmissivity and local slope (O'Loughlin, 1986). Flow is assumed to infiltrate to a lower conductivity layer and follow topographically determined flow paths. Local wetness (W) is calculated as the ratio of the local flux at a given steady-state rainfall (Q) to that upon complete saturation of the soil profile

$$W = \frac{Qa}{bT \sin\theta} \tag{2}$$

where a is the upslope contributing area (m^2), b is the length across which flow is accounted for (m), T is the depth-integrated soil transmissivity (m^2/day) and θ is the local slope (degrees). Adopting the simplifying assumption that the saturated conductivity does not vary with depth, Equation (1) can be reduced for the case where $W \leqslant 1$ to

$$W = h/z \tag{2}$$

where h is the thickness of the saturated soil above the impermeable layer and z is the total thickness of the soil. Combining Equations (1) and (2) allows expression of the relative saturation of the soil profile as

$$h/z = \frac{Qa}{bT \sin\theta} \tag{3}$$

which predicts relative soil moisture as a function of steady-state rainfall, specific catchment area (a/b), soil transmissivity and local slope.

The infinite-slope stability model provides a one-dimensional model for failure of shallow soils that neglects arching and lateral root reinforcement. Under these assumptions the criterion for slope failure may be expressed as

$$\rho_s g z \sin\theta \cos\theta = C' + [\rho_s - (h/z)\rho_w]gz \cos^2\theta \tan\phi \tag{4}$$

where ρ_s is the bulk density of the soil, g is gravitational acceleration, z is soil thickness, C' is the effective cohesion of the soil including the effect of reinforcement by roots that penetrate the basal failure surface, ρ_w

is the bulk density of water and ϕ is the friction angle of the soil (e.g. Selby, 1982). Combining Equations (3) and (4) and rearranging in terms of the critical steady-state rainfall (Q_c) necessary to trigger slope instability, yields

$$Q_c = \frac{T \sin\theta}{(a/b)} \left[\frac{C'}{\rho_w g z \cos^2\theta \tan\phi} + \frac{\rho_s}{\rho_w} \left(1 - \frac{\tan\theta}{\tan\phi} \right) \right] \qquad (5a)$$

For the case of cohesionless soils (i.e. $C' = 0$) this reduces to the model explored in greater detail by Montgomery and Dietrich (1994)

$$Q_c = \frac{T \sin\theta}{(a/b)} \left[\frac{\rho_s}{\rho_w} \left(1 - \frac{\tan\theta}{\tan\phi} \right) \right] \qquad (5b)$$

The physical interpretation of W greater than 1·0 is that for the excess water runs off as overland flow; hence, there is no mechanism in this model for generating pore pressures in excess of hydrostatic pressures. Slopes that are stable even when $W = 1·0$ are interpreted to be unconditionally stable and to require excess pore pressures to generate slope instability. The condition for unconditionally stable slopes can be expressed as

$$\tan\theta < \frac{C'}{\rho_s g z \cos^2\theta} + \left(1 - \frac{\rho_w}{\rho_s} \right) \tan\phi \qquad (6)$$

Similarly, slopes predicted to be unstable even when dry (i.e. $W = 0$) are considered to be unconditionally unstable areas where soil accumulation would be difficult and therefore one would expect to find rock outcropping. The condition for unconditionally unstable slopes can be expressed as

$$\tan\theta \geqslant \tan\phi + \frac{C'}{\rho_s g z \cos^2\theta} \qquad (7)$$

Critical rainfall values can be calculated for locations with slopes between the criteria of Equations (6) and (7).

The model given by Equations (5)–(7) can be implemented using either topographic elements defined by contours and flow lines, or by the square grid cells typical of digital elevation models (DEMs). Comparison of the location of field-mapped landslides and Q_c values predicted under the assumption of cohesionless soils supports the interpretation that locations with equal Q_c have equal topographic control on shallow landslide initiation in three small drainage basins in the western United States (Montgomery and Dietrich, 1994). The steady-state rainfall and cohesionless soil assumptions, however, complicate translation of Q_c values into actual failure probabilities. Lateral reinforcement by roots that extend across the side of potential failures (Burroughs, 1984; Reneau and Dietrich, 1987) and systematic variations in soil thickness also influence the probability of slope failure (Dietrich et al., 1995). Without calibration to field data our approach can only identify areas with equal topographic control on shallow landslide initiation.

Spatial variations in soil properties also influence the probability of failure and the location of specific failures may be strongly influenced by site-specific details such as interaction of flow in colluvial soil and near-surface fractured bedrock (e.g. Montgomery et al., 1997). Variation in the amount of hydrological leakage to deeper groundwater flow may also strongly influence patterns of landsliding. Consequently, it is likely that within areas of equal topographic control on shallow landslide initiation some locations will be more or less susceptible to failure. Moreover, soil thickness increases through time in topographic hollows (e.g. Dietrich and Dunne, 1978; Dietrich et al., 1995) and these changes will lead to an increasing probability of slope failure as a hollow infills with colluvium (Dunne, 1991; Montgomery, 1991). At a specific location, however, the other soil and topographic properties incorporated in Equation (5) do not change through time and failures are initiated through changes in the thickness of the saturated zone in response to high intensity or long duration storms (Caine, 1980).

The magnitude of soil reinforcement by root strength varies both between species and in response to timber felling (Takahasi, 1968; Endo and Tsuruta, 1969; Burroughs and Thomas, 1977; Ziemer and Swanston, 1977; Wu et al., 1979; Riestenberg and Sovonick-Dunford, 1983; Reneau and Dietrich, 1987). Much of the total tensile strength of root systems resides in the finest roots, which die back rapidly following forest clearing (Burroughs and Thomas, 1977; Ziemer and Swanston, 1977). Total root strength decreases to a minimum between 3 and 15 years after timber harvest and returns to values similar to mature forest only after several decades (e.g. Sidle, 1992). Burroughs and Thomas (1977) estimated total tensile strength per unit area of soil imparted by mature Coast Douglas fir (*Psuedotsuga menziesii*) to be about 17 kN/m^2, declining rapidly after cutting to about 2 kN/m^2 for stumps. Less cohesion is attributable to root systems of other species; hardwood species range from 2 to 13 kN/m^2 (Takahasi, 1968; Endo and Tsuruta, 1969; Riestenberg and Sovonick-Dunford, 1983; Reneau and Dietrich, 1987); woody shrubs and ground cover are typically <3 kN/m^2 (Burroughs, 1984; Terwilliger and Waldron, 1991). The effective cohesion imparted to a soil by tree roots varies spatially, with depth and spacing of trees and with tree age. Although the full tensile strength of roots is not always mobilized during slope failure, examination of debris flow scarps in California, Oregon and Washington revealed that most exposed roots were sheared in the lateral margins and headscarps of shallow landslides. Comparison of the net root strength required for slope stability under the infinite-slope model and a lateral root strength model (Reneau and Dietrich, 1987) reveals that significantly greater root strength is required for the finite (i.e. lateral) than the infinite-slope model. Hence, calculations based on the infinite-slope model should yield a minimum constraint on the contribution of root strength to soils in which roots do not extend through the basal failure surface.

STUDY AREAS AND METHODS

We analysed data from 14 watersheds assessed under the Washington State watershed analysis methodology (WFPB, 1993). The study watersheds range in size from 119 km^2 to 323 km^2 and occupy a total of 2993 km^2 spread across several physiographic provinces (Figure 1) and encompassing a wide range of geology and landforms (Table I). Each of the watersheds has been completely logged over the past 100 years, and some areas are in their third rotation of industrial forestry.

We chose to model relative slope stability using a single set of soil properties in order to focus on the topographic control on shallow landsliding. Based on extensive field measurement at a study site in coastal Oregon (Montgomery et al., 1997) we used values of $T = 65$ m^2/day, $\rho_s = 2000$ kg/m^3, $z = 1.0$ m and $\phi = 33°$ for our analyses of each watershed, although we recognize that local variations in geology and soils should strongly influence patterns of landsliding. We therefore expect the model to perform poorly in areas with high recharge to deep groundwater, and other areas where the model assumptions are not well met. We presume that models which incorporate spatially distributed, watershed-specific soil and geological factors should perform better than the general model we investigate here. The use of a single set of parameter values, however, allows a unique regional test of the influence of topographic forcing on shallow landsliding, and also allows for comparison of the effect of lithology by controlling for topographic influences.

Landslides visible on aerial photographs were mapped on to 1:24000 scale topographic maps in each watershed. Aerial photograph coverage extended from the 1940s to the 1990s in most of the watersheds. Landslides were classified into shallow and deep-seated varieties, and the former were subdivided into 'in unit' failures and those interpreted to reflect road-related influences (e.g. drainage concentration, fill berm failures, etc.). Landslide mapping did not always delineate slide scarps separately; slide outlines therefore may include initiation, transport and depositional areas. Also, comparison of aerial photographs and composite slide maps indicates that many small slides were not mapped in the watershed analyses apparently due to time, budget and resolution constraints. Hence, the extent of landsliding revealed by the watershed analyses provides a minimum estimate of contemporary rates of shallow landsliding.

Figure 1. Location of the study watersheds

Table I. Characteristics of watersheds included in the study area

State	Watershed	Physiographical area	Dominant lithology	Area (km²)
Washington	Griffin/Tokul	Puget Sound Lowlands	Glacial outwash	159·1
	Tolt	Puget Sound Lowlands/ Central Cascades	Glacial outwash/andesites	244·5
	Skookumchuck	Central Cascades	Basalt/andesite	158·9
	Coweeman	Central Cascades	Basalt/andesite	182·3
	Stillman	Coast Range	Tertiary sedimentary/extrusive and intrusive basalt	118·5
	Chehalis	Coast Range	Tertiary sedimentary/extrusive and intrusive basalt	181·7
	Vesta/Little North	Coast Range	Sandstone/siltstone	227·3
	Willapa	Coast Range	Sedimentary/extrusive and intrusive basalt	254·7
Oregon	Siuslaw	Coast Range	Sedimentary sandstone and siltstone	270·6
	Millicoma	Coast Range	Sedimentary sandstone and siltstone	230·9
	Williams	Coast Range	Sedimentary sandstone and siltstone	322·9
	Mohawk	Cascades foothills	Pyroclastic	211·2
	Lower McKenzie (north)	Cascades	Tertiary volcanics	244·2
	Lower McKenzie (south)	Cascades	Tertiary volcanics	206·6
Total				2993·4

The outline of each landslide was digitized to create a landslide polygon coverage for GIS analysis. The digitized landslide layer was rectified to composite 30 m grid USGS digital elevation models of each catchment. Critical rainfall values were calculated using the parameter values discussed above. Each landslide was attributed to the lowest Q_c category that it overlapped in order to account for both mapping and registration errors and the inclusion of transport and depositional areas in the slide polygons. Although simulations were run for three different cohesion values for each watershed (2, 8 and 15 kN/m^2), here we focus on the analyses for $C' = 2$ kN/m^2, which approximates the post-cutting root strength minima in this region (Figure 2). Comparison of the relative frequency of slides within each Q_c category provides for direct testing of the assumption that lower Q_c implies higher failure frequency.

RESULTS

The distribution of the 3224 mapped landslides indicates that rates of shallow landsliding varied dramatically between the watersheds (Table II). The Skookumchuck watershed had the lowest number of observed slides (39) and the Chehalis watershed had the highest (629). Grouping of landslide frequency data by physiographic provinces yields subregional trends (Figure 3). Watersheds in coastal Washington underlain by sedimentary and igneous rocks have the highest rates of sliding; rates are substantially lower in the other areas.

In spite of such differences in slide frequency, the size distribution of shallow landslides mapped in these watersheds follows a roughly exponential distribution (Figure 4). Very few slides involve more than 20 000 m^2 and most involve less than 10 000 m^2; the mean slide size was 8152 m^2 and the median size was 3386 m^2. Because of their great number, small slides contribute substantially to the net sediment flux. The sizes of the mapped landslide polygons, however, reflect both mapping scale and the fact that the size of the total area influenced by the slide can be much larger than the initiation zone. The initial area of typical shallow landslides in this region is less than several thousand square metres (Pierson, 1977; Reneau and Dietrich, 1987), and the exponential distribution of total slide sizes quantifies the addition of material scoured from downstream areas in the runout path. Although it appears that a simple exponential function may account for variations in the downslope growth of slides, the disproportionately large number of slides

Table II. Landslide frequency (slides/km^2) in each critical rainfall category (mm/day) for the 14 study watersheds based on $C' = 2$ kN/m^2 and other parameters cited in the text

Watershed	Uncond. unstable	0–50	50–100	100–200	200–400	400 +	Uncond. stable	Watershed Total
Chehalis	30·5	19·4	8·5	4·5	2·7	0·9	0·5	3·5
Coweeman	6·5	10·6	3·1	0·8	0·5	0	0·2	0·5
Griffin/Tokul	0	7·6	3·6	1·1	1·4	1·9	0·5	0·5
Millicoma	8·9	3·7	2·3	1·8	1·6	2·5	0·3	1·2
Mohawk	3·7	0·8	0·8	0·4	0·3	0	<0·1	0·2
N. McKenzie	19·0	4·0	1·0	0·3	0	1·8	0·1	1·4
S. McKenzie	12·8	2·3	1·1	0·6	0·4	0	0·1	0·9
Siuslaw	7·3	2·8	1·5	1·2	0·9	4·0	0·1	0·4
Skookumchuck	4·9	2·6	0·5	0·5	0·3	0	<0·1	0·2
Stillman	0	16·2	9·9	7·4	2·6	3·0	0·7	1·4
Tolt	1·3	2·2	1·2	1·0	1·1	0	0·4	0·7
Vesta	0	30·6	30·8	19·0	10·2	4·4	0·7	1·3
Willapa	35·9	20·3	8·9	3·9	2·0	1·0	0·5	1·3
Williams	6·7	3·0	3·1	2·5	2·1	0·9	0·3	1·4
Totals	9·7	4·5	2·9	2·5	2·2	1·2	0·3	1·1

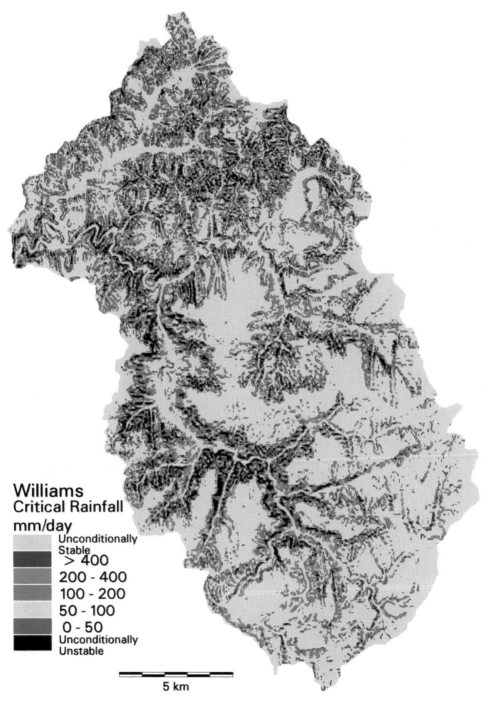

Williams
Critical Rainfall
mm/day

Unconditionally
Stable
> 400
200 - 400
100 - 200
50 - 100
0 - 50
Unconditionally
Unstable

5 km

Figure 2(a)

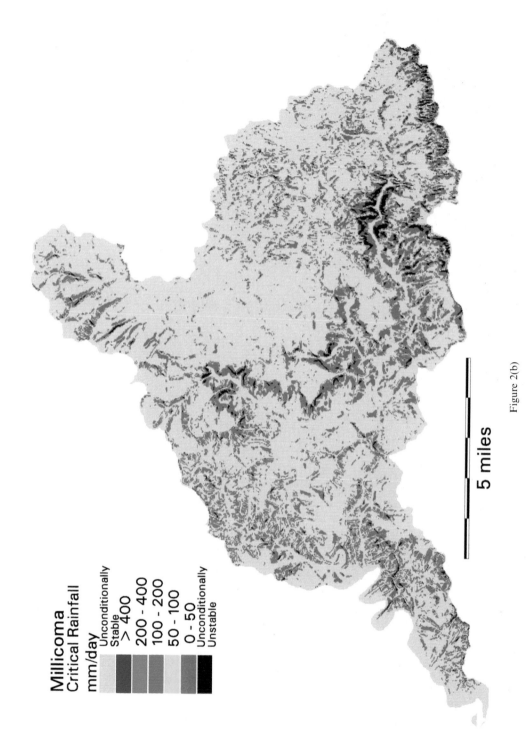

Millicoma
Critical Rainfall
mm/day

Unconditionally
Stable
> 400
200 - 400
100 - 200
50 - 100
0 - 50
Unconditionally
Unstable

5 miles

Figure 2(b)

Figure 2(a–b). Maps illustrating the distribution of Q_c values for $C = 2$ kN/m^2 in the Williams and Millicoma watersheds

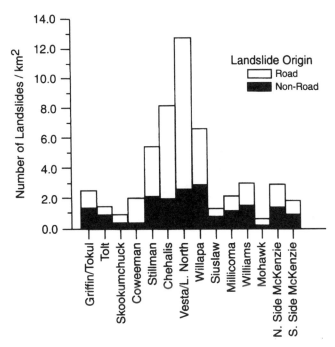

Figure 3. Aggregate landslide frequency for each watershed. Note that the coastal Washington watersheds have the highest rates of sliding

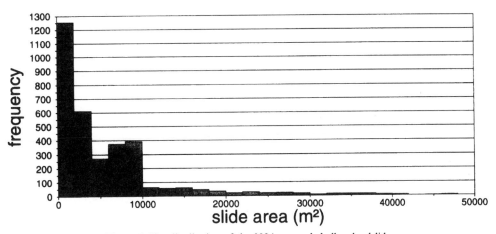

Figure 4. Size distribution of the 3224 mapped shallow landslides

about 10 000 m^2 in size may record enhanced growth of slides favourably oriented for propagation through the channel network (e.g. Grant *et al.*, 1984; Benda and Dunne, 1987; Benda, 1990; Benda and Cundy, 1990).

The number of slides associated with each Q_c category varied widely between the watersheds. In some of the basins the majority of the slides occurred in the lowest Q_c categories (e.g. N. McKenzie), whereas in one basin (i.e. Griffin/Tokul) most slides occurred in areas predicted to be stable. The proportion of slides that occurred within Q_c categories <100 mm/day ranged from 7 to 92% across the study watersheds, and the proportion of slides that occurred within Q_c categories <200 mm/day ranged from 9 to 93%. The frequency of slides per unit area, however, consistently showed greater rates of sliding in areas predicted to have lower Q_c both within each watershed (Table II) and across all of the watersheds (Figure 5).

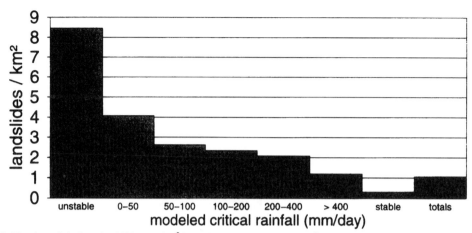

Figure 5. Number of shallow landslides per km² in each Q_c category for the combined data set from all 14 study watersheds

We used a non-parametric test for population similarity to evaluate the hypothesis that landslides occur disproportionately in areas of low Q_c. The null hypothesis is that the distribution of landslides should occur in direct proportion to the area mapped in each Q_c category if the model does not discriminate relative landslide hazard. Equality of the two populations was tested with Pearson's χ^2 statistic for goodness-of-fit for categorized data (Bhattacharyya and Johnson, 1977). The test statistic is computed as

$$\chi^2 = \Sigma \frac{(O - E)^2}{E} \tag{8}$$

where O is the observed frequency of landslides in each Q_c category and E is the expected number calculated as the proportion of total landslides based on the area of the watershed in the Q_c category. The test statistic was statistically significant for each watershed (Table III), indicating rejection of the null hypothesis and implying that the model discriminates areas of greater landslide hazard. Hence, Q_c provides a surrogate for failure initiation probability as a function of topographic position, as hypothesized by Montgomery and Dietrich (1994).

Counter to the conventional wisdom that roads cause the vast majority of shallow slides in the Pacific Northwest, the 1881 road-related slides account for just over half (58%) of the mapped slides. In most of the watersheds, road-related slides and 'in-unit' slides occur with roughly equal frequency; basins with many road-related slides also have many 'in-unit' slides (Figure 6). However, in two of the watersheds (Chehalis and Vesta/Little North) road-related slides are far more numerous than 'in-unit' slides. Least-squares linear regression of the total number of 'in-unit' slides against the total number of road-related slides for all but these two watersheds yields a relationship of $y = 8 + 0.97x$ ($R^2 = 0.85$), which indicates that road-related and 'in-unit' failure rates account for a comparable number of landslides in the majority of the watersheds. Hence, 'in-unit' failures appear to account for about half the slides on a regional basis, although roads can be responsible for the vast majority of shallow landsliding in particular watersheds. Perhaps even more surprising is that the relationship between Q_c and the relative frequency of road-related failures parallels that for 'in-unit' slides. Road-related slides are concentrated in areas of low Q_c; topographic control still seems to dominate the location of road-related slides, which appear to trigger slides in areas prone to failure.

Consistent patterns in watershed sensitivity to reduced root strength can be shown by comparing the predicted extent of potentially unstable ground for low root strength versus that for high root strength. In all of the watersheds, simulations with root strengths of 8 and 16 kN/m² virtually eliminated areas of potential instability identified in the 2 kN/m² simulations. For the higher root strength cases, incorporation of a greater soil depth as typically found in topographic hollows, localizes the zones of predicted potential instability in hollows (Dietrich et al., 1995). Hence, a dramatic increase in the rate of shallow landsliding

Table III. χ^2 test statistic for the Pearson goodness-of-fit for categorized data for landslide frequency in Q_c categories modelled at cohesion $= 2$ kN/m^2. Test statistic greater than 18 is significant at $P < 0.005$ (at 6 d.f.)

Watershed	χ statistic
Griffin/Tokul	311
Tolt	163
Skookumchuck	195
Coweeman	564
Stillman	420
Chehalis	1577
Vesta/L. North	1562
Willapa	2308
Siuslaw	249
Millicoma	382
Mohawk	132
N. McKenzie	967
S. McKenzie	610

should be expected to result from loss of root strength following forest clearance in these steep, soil-mantled basins.

DISCUSSION AND CONCLUSIONS

The strong correlation between rates of sliding and calculated Q_c indicates that topographic control dominates the location of shallow landslides in Pacific Northwest watersheds. If local geology or site-specific soil or hydrological conditions dominate the distribution of shallow landsliding, then there should be no such pattern and the model should perform poorly. Instead, the model performs surprisingly well given the use of a single set of parameters across a wide range of lithology and topographic settings. Consequently, we interpret our findings as strong support for considering topographically driven convergence of near-surface runoff as a primary control on shallow landsliding in this region, as has been long-advocated by Dietrich and colleagues (e.g. Dietrich *et al.*, 1982, 1986; Reneau and Dietrich, 1987; Montgomery and Dietrich, 1988).

The importance of topographically driven slide frequency highlights the relationship between shallow landsliding and landscape evolution, as more frequent failure implies more rapid erosion. We interpret our analyses as strong support for the view that shallow landsliding plays a primary role in the incision of headwater valleys in steep terrain, as advocated by Seidl and Dietrich (1992). The inverse exponential form of the relationship between slide frequency and Q_c also provides insight into the construction of a sediment transport law for shallow landsliding. Many landscape evolution models consider sediment flux by landsliding to be a diffusive, slope-dependent process and hence that landslides act to smooth rather than create local relief. In contrast, considering sediment flux by shallow landslides (Q_s) to depend upon Q_c (i.e. $Q_s = aQ_c^{-b}$) would treat landsliding as incisive by including both a slope and an area dependence.

Two types of model error are possible: in type I errors the model predicts that landslides are not likely to occur but they do, whereas in type II errors the model predicts that landslides are likely to occur but they do not. Each watershed has far more area mapped in low Q_c than has actually experienced slope failure. We interpret the model to provide a relative likelihood of failure that can be considered to integrate a stochastic process of failure initiation across time and space. Hence, we consider unfailed areas with low Q_c to delineate likely sites of failure in future storms rather than areas presumed to have been tested and proved stable during previous storms.

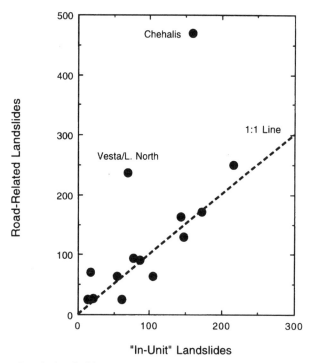

Figure 6. Number of road-related slides versus number of non-road-related slides in each of the study watersheds

A total of 24% of the mapped landslides occur in areas that the model predicts to be unconditionally stable. Road-related effects account for almost half of these failures and the proportion of such type I errors varied greatly between watersheds. As few as 6% of the landslides were missed by the model in several watersheds and 88% of the mapped slides occurred in 'stable' ground in the watershed with the worst model performance (Griffin/Tokul). Many of the type I errors arose from the action of processes external to the model framework and local conditions specific to the study watersheds. The model generally performed well in watersheds underlain by shallow deposits and poorly in areas with thick glacial deposits (e.g. Griffin/ Tokul) where groundwater recharge may control soil saturation and shallow landsliding. Field work in many of the areas where shallow landslides occurred in locations predicted to be unconditionally stable revealed that many of the slides occurred: (i) in steep inner gorges not resolved in the 30 m DEMs; (ii) from the scarps, flanks and toes of large, deep-seated landslides; and (iii) in association with other local effects such as 30 m high abandoned logging railroad embankments that do not appear on either 7·5′ topographic maps or on 30 m DEMs (P. Russel, personal communication). Such factors external to the model framework can reasonably account for the < 10% of the mapped landslides that remain inexplicable within the context of the model.

The influence of roads and lithology on shallow landslide frequency holds several surprising implications. The similar relationship between slide frequency and Q_c for road-related and in-unit slides indicates that the acceleration of landsliding owing to road construction appears to reflect where they are built as much as how they are built. Although the style of road construction can profoundly influence landslide hazards (e.g. Furniss *et al.*, 1991), siting of roads across, or the delivery of road drainage to, potentially unstable ground also has a primary influence on landslide frequency. Our data also confirm that differences in geology impose a broad control on absolute rates of shallow landsliding upon which superimposed topographic controls dominate the relative slide frequency. This suggests the wisdom of dividing large basins or regions into areas of similar geology in order to calibrate modelling of sediment flux or failure rates. Such a larger scale geological context is particularly important for interpreting differences in slide frequencies between different watersheds.

The inherent susceptibility of a landscape to changes in C' that accompany timber harvest varies dramatically, both within a watershed and regionally. At present, however, relatively little use is made of the differences in landscape susceptibility to timber harvesting in land use planning. Instead, ownership patterns tend to drive definitions of suitable land use independent of the relative risk inherent in the different terrains. Models such as those investigated here could provide useful tools for addressing the inherent capacity of the landscape to sustain particular land uses, a key issue in the design and evaluation of watershed management strategies (e.g. Montgomery *et al.*, 1995).

A key implication of our analyses is that high C', as characterizes mature coniferous forest in the Pacific Northwest, effectively limits shallow landsliding to areas with deep soils, such as hollows that fill with colluvium, or where local hydrological conditions lead to excess pore pressures. Processes that lower C' (e.g. fire, timber harvesting) should lead to higher probabilities of failure across a greater proportion of the landscape.

Our analysis further motivates discussion of appropriate roles for models in watershed planning. In particular, our data document that Q_c provides a reasonable surrogate for failure probability after timber harvest and that most slides occur from readily identifiable areas in which root strength appears critical to slope stability. Establishing appropriate bounds on what range of Q_c values should be considered hazardous is an important question for use of the model in landslide hazard assessment. One approach is to determine an acceptable level of slide protection (e.g. say to minimize the risk from 80% of slides) for a watershed and then use the observed landslide record to predict the Q_c value that would impose appropriate protection across the watershed. The appropriate level of protection and the specific land management prescriptions to provide that protection are policy questions that require consideration of risk as well as hazard and a clear set of land use objectives and priorities. Models such as those investigated here can inform development of landslide hazard mitigation measures and provide a theoretical framework for the evaluation and interpretation of both assessment and monitoring of data and the effectiveness of land management prescriptions.

ACKNOWLEDGEMENTS

This study was supported by the Weyerhaeuser Company, the Cooperative State Research Service, US Department of Agriculture (No. 94-37101-0321), US Forest Service (co-op agreement PNW 93-0441), NSF (grant CMS-9610269) and a gift from the Washington Forest Protection Association. Perianne Russel mapped many of the landslides during watershed analyses and provided insight into interpretation of model performance. Numerous conversations with Bill Dietrich have greatly influenced our thinking on the use and interpretation of the type of model examined here.

REFERENCES

Benda, L. 1990. 'The influence of debris flows on channels and valley floors of the Oregon Coast Range, USA.', *Earth Surf. Process. Landf.*, **15**, 457–466.

Benda, L. and Cundy, T. W. 1990. 'Predicting deposition of debris flows in mountain channels', *Can. Geotech. J.*, **27**, 409–417.

Benda, L. and Dunne, T. 1987. 'Sediment routing by debris flow', in Beschta, R. L., Blinn, T., Grant, G. E., Swanson, F. J., and Ice, G. C. (eds), *Erosion and Sedimentation in the Pacific Rim, IAHS Publ.*, **165**, 213–223.

Bhattacharyya, G. K. and Johnson, R. A. 1977. *Statistical Concepts and Methods*. John Wiley & Sons, New York, 639 pp.

Brabb, E. E., Pampeyan, E. H., and Bonilla, M. G. 1972. 'Landslide susceptibility in San Mateo County, California, scale 1:62 500', *Miscellaneous Field Studies Map, MF-360*. US Geological Survey, Washington, D.C.

Brown, G. W. and Krygier, J. T. 1971. 'Clear-cut logging and sediment production in the Oregon Coast Range', *Wat. Resour. Res.*, 7, 1189–1198.

Burroughs, E. R. Jr. 1984. 'Landslide hazard rating for portions of the Oregon Coast Range', in O'Loughlin, C. L., and Pearce, A. J. (eds.), *Proceedings of the Symposium on Effects of Forest Land Use on Erosion and Slope Stability, Honolulu, Hawaii*, Environment and Policy Institute, University of Hawaii, Honolulu. pp. 265–274.

Burroughs, E. R. Jr and Thomas, B. R. 1977. 'Declining root strength in Douglas-Fir after felling as a factor in slope stability', *Forest Service Research Paper INT-190*. US Department of Agriculture, Ogden. 27 pp.

Busoni, E., Sanchis, P. S., Calzolari, C., and Romognoli, A. 1995. 'Mass movement and erosion hazard patterns by multivariate analysis of landscape integrated data: the Upper Orcia River Valley (Siena, Italy) case', *Catena*, **25**, 169–185.

Caine, N. 1980. 'The rainfall intensity–duration control of shallow landslides and debris flows', *Geograf. Annal., Ser. A*, **62**, 23–27.

Campbell, R. H. 1975. 'Soil slips, debris flows and rainstorms in the Santa Monica Mountains and vicinity, southern California', *US Geological Survey Professional Paper 851*. US Geological Survey, Washington, D.C. 51 pp.

Carrara, A. 1983. 'Multivariate models for landslide hazard evaluation', *Math. Geol.*, **15**, 403–426.

Carrara, A., Pugliese Carratelli, E., and Merenda, L. 1977. 'Computer-based bank and statistical analysis of slope instability phenomena', *Z. Geomorph.*, **21**, 187–222.

Carrara, A., Cardinali, M., Detti, R., Guzzetti, F., Pasqui, V., and Reichenback, P. 1991. 'GIS techniques and statistical models in evaluating landslide hazard', *Earth Surf. Process. Landf.*, **16**, 427–445.

Derbyshire, E., van Asch, T., Billard, A., and Meng, X. 1995. 'Modelling the erosional susceptibility of landslide catchments in thick loess: Chinese variations on a theme by Jan de Ploey', *Catena*, **25**, 315–331.

Dietrich, W. E. and Dunne, T. 1978. 'Sediment budget for a small catchment in mountainous terrain', *Z. Geomorph.*, **29** (Suppl.), 191–206.

Dietrich, W. E., Dunne, T., Humphrey, N., and Reid, L. 1982. 'Construction of sediment budgets for drainage basins', in Swanson, F. J., Janda, R. J., Dunne, T., and Swanston, D. N. (eds), *Sediment Budgets and Routing in Forested Drainage Basins*, US Department of Agriculture, Forest Service General Technical Report PNW-141. Pacific Northwest Forest and Range Experiment Station, Portland, Oregon. pp. 2–23.

Dietrich, W. E., Wilson, C. J., and Reneau, S. L. 1986. 'Hollows, colluvium, and landslides in soil-mantled landscapes', in Abrahams, A. D. (ed.), *Hillslope Processes*. Allen and Unwin, Boston. pp. 361–388.

Dietrich, W. E., Wilson, C. J., Montgomery, D. R., and McKean, J. 1993. 'Analysis of erosion thresholds, channel networks and landscape morphology using a digital terrain model', *J. Geol.*, **101**, 259–278.

Dietrich, W. E., Reiss, R., Hsu, M.-L., and Montgomery, D. R. 1995. 'A process-based model for colluvial soil depth and shallow landsliding using digital elevation data', *Hydrol. Process.*, **9**, 383–400.

Dunne, T. 1991. 'Stochastic aspects of the relations between climate, hydrology and landform evolution', *Trans. Jpn. Geomorph. Union*, **12**, 1–24.

Endo, T. and Tsuruta, T. 1969. 'The effect of tree roots on the shearing strength of soil', *Annual Report*. Forest Experiment Station, Hokkaido. pp. 167–182.

Fredriksen, R. L. 1970. 'Erosion and sedimentation following road construction and timber harvest on unstable soils in three small western Oregon watersheds', *Forest Service Research Paper PNW-104. US Department of Agriculture*. Porland. 15 p.

Furniss, M. J., Roelofs, T. D., and Yee, C. S. 1991. 'Road construction and maintenance', in Meehan, W. R. (ed.), *Influences of Forest and Rangeland Management on Salmonid Fishes and Their Habitats. Am. Fish. Soc. Spec. Publ.*, **19**, 297–323.

Grant, G. E., Crozier, M. J., and Swanson, F. J. 1984. 'An approach to evaluating off-site effects of timber harvest activities on channel morphology', in *Proceedings, Symposium on the Effects of Forest and Land Use on Erosion and Slope Stability, Environment and Policy Institute, East-West Center, University of Hawaii, Honolulu*. pp. 177–186.

Gresswell, S., Heller, D., and Swanston, D. N. 1979. 'Mass movement response to forest management in the central Oregon Coast Ranges', *Forest Service Resources Bulletin PNW-84*, US Department of Agriculture, Portland. 26 pp.

Hollingsworth, R. and Kovacs, G. S. 1981. 'Soil slumps and debris flows: prediction and protection', *Bull. Assoc. Engng. Geol.*, **18**, 17–28.

Lanyon, L. E. and Hall, G. F. 1983. 'Land-surface morphology: 2. Predicting potential landscape instability in eastern Ohio', *Soil Sci.*, **136**, 382–386.

Mark, R. K. 1992. 'Map of debris-flow probability, San Mateo County, California, scale 1:62 500, *Miscellaneous Investigations Map, I-1257-M*. US Geological Survey, Washington, D.C.

Martin, K., Skaugset, A., and Pyles, M. R. 1996. 'Forest management of landslide prone sites: the effectiveness of headwall leave areas, part II', *COPE Report*, **9**, 8–12.

Mersereau, R. C. and Dyrness, C. T. 1972. 'Accelerated mass wasting after logging and slash burning in western Oregon', *J. Soil Wat. Conserv.*, **27**, 112–114.

Montgomery, D. R. 1991. 'Channel initiation and landscape evolution', *PhD Thesis*, University of California, Berkeley. 421 pp.

Montgomery, D. R. and Dietrich, W. E. 1988. 'Where do channels begin?', *Nature*, **336**, 232–234.

Montgomery, D. R. and Dietrich, W. E. 1994. 'A physically-based model for the topographic control on shallow landsliding', *Wat. Resour. Res.*, **30**, 1153–1171.

Montgomery, D. R., Wright, R. H., and Booth, T. 1991. 'Debris flow hazard migration for colluvium-filled swales', *Bull. Assoc. Engng. Geol.*, **28**, 299–319.

Montgomery, D. R., Grant, G. E., and Sullivan, K. 1995. 'Watershed analysis as a framework for implementing ecosystem management', *Wat. Resour. Bull.*, **31**, 369–386.

Montgomery, D. R., Dietrich, W. E., Torres, R., Anderson, S. P., Heffner, J. T., and Loague, K. 1997. 'Hydrologic response of a steep unchanneled valley to natural and applied rainfall', *Wat. Resour. Res.*, **33**, 91–109.

Neuland, H. 1976. 'A prediction model of landslips', *Catena*, **3**, 215–230.

Neuland, H. 1980. 'Diskriminanzanalytische untersuchungen zur identifikation der auslösefaktoren für rutschungen in verschiedenen höhenstuufen der Kolumbianischen Anden', *Catena*, **7**, 205–221.

Niemann, K. O. and Howes, D. E. 1991. 'Applicability of digital terrain models for slope stability assessment', *ITC J.*, **1991-3**, 127–137.

Okimura, T. and Ichikawa, R. A. 1985. 'Prediction method for surface failures by movements of infiltrated water in a surface soil layer', *Natural Disaster Sci.*, **7**, 41–51.

Okimura, T. and Nakagawa, M. 1988. 'A method for predicting surface mountain slope failure with a digital landform model', *Shin Sabo*, **41**, 48–56.

O'Loughlin, E. M. 1986. 'Prediction of surface saturation zones in natural catchments by topographic analysis', *Wat. Resour. Res.*, **22**, 794–804.

Pierson, T. C. 1977. 'Factors controlling debris-flow initiation on forested hillslopes in the Oregon Coast Range', *PhD Thesis*, University of Washington, Seattle. 166 pp.

Reneau, S. L. and Dietrich, W. E. 1987. 'Size and location of colluvial landslides in a steep forested landscape', in Beschta, R. L., Blinn, T., Grant, G. E., Ice, G. G., and Swanson, F. J. (eds), *Proceedings of the International Symposium on Erosion and Sedimentation in the Pacific Rim, IAHS Pub.*, **165**, 39–48.

Riestenberg, M. M. and Sovonick-Dunford, S. 1983. 'The role of woody vegetation in stabilizing slopes in the Cincinnati area, Ohio', *Geol. Soc. Am. Bull.*, **94**, 506–518.

Seeley, M. W. and West, D. O. 1990. 'Approach to geologic hazard zoning for regional planning, Inyo National Forest, California and Nevada', *Bull. Assoc. Eng. Geol.*, **27**, 23–35.

Seidl, M. and Dietrich, W. E. 1992. 'The problem of channel incision into bedrock', in Schmidt, K.-H. and de Poley, J. (eds), *Functional Geomorphology*, Catena Supplement 23. Catena Verlag, Cremlingen. pp. 101–124.

Selby, M. J. 1982. *Hillslope Materials and Processes*. Oxford University Press, Oxford. 264 pp.

Sidle, R. C. 1992. 'A theoretical model of the effects of timber harvesting on slope stability', *Wat. Resour. Res.*, **28**, 1897–1910.

Skaugset, A., Froehlich, H., and Lautz, K. 1993. 'The effectiveness of headwall leave areas', *COPE Report*, **6**, 3–6.

Swanson, F. J., and Dyrness, C. T. 1975. 'Impact of clear-cutting and road construction of soil erosion by landslides in the western Cascade Range, Oregon', *Geology*, **3**, 393–396.

Swanson, F. J. Fredriksen, R. L., and McCorison, F. M. 1982. 'Material transfer in a western Oregon forested watershed', in Edmonds, R. L. (ed.), *Analysis of Coniferous Forest Ecosystems in the Western United States*. Hutchinson Ross Publishing Co., Stroudsburg. pp. 233–266.

Swanston, D. N. and Swanson, F. J. 1976. 'Timber harvesting, mass erosion, and steepland forest geomorphology in the Pacific Northwest', in Coates, D. R. (ed.), *Geomorphology and Engineering*. Dowden, Hutchinson & Ross, Inc., Stroudsburg. pp. 199–221.

Takahasi, T. 1968. 'Studies of the forest facilities to prevent landslides', *Bull Fac. Agric., Shizuoka Univ.*, **18**, 85–101.

Terwilliger, V. J. and Waldron, L. J. 1991. 'Effects of root reinforcement on soil-slip patterns in the Transverse Ranges of southern California', *Geol. Soc. Am. Bull.*, **103**, 775–785.

van Asch, T., Kuipers, B., and van der Zanden, D. J. 1993. 'An information system for large scale quantitative hazard analyses of landslides', *Z. Geomorph.*, **87** (Suppl.), 133–140.

WFPB (Washington Forest Practice Board), 1993. *Standard Methodology for Conducting Watershed Analysis, Version 2·0*. WFPB, 85 pp.

Williams, G. P. and Guy, H. P. 1971. 'Debris avalanches — a geomorphic hazard', in Coates, D. R. (ed.), *Environmental Geomorphology*. SUNY Publications in Geomorphology, Binghampton. pp. 25–46.

Wu, W. and Sidle, R. C. 1995. 'A distributed slope stability model for steep forested basins', *Wat. Resour. Res.*, **31**, 2097–2110.

Wu, T. H., McKinnell, W. P., III, and Swanston, D. N. 1979. 'Strength of tree roots and landslides on Prince of Wales Island, Alaska', *Can. Geotech. J.*, **16**, 19–33.

Ziemer, R. R. and Swanston, D. N. 1977. 'Root strength changes after logging in southeast Alaska', *Research Note PNW-306*. US Forest Service, Pacific Northwest Forest and Range Experiment Station, Portland.

10

REGIONAL-SCALE ASSESSMENT OF NON-POINT SOURCE GROUNDWATER CONTAMINATION

KEITH LOAGUE[1]* AND DENNIS L. CORWIN[2]

[1]*Department of Geological and Environmental Sciences, Stanford University, Stanford, CA 94305-2115, USA*
[2]*US Salinity Laboratory, USDA–ARS, 450 West Big Springs Road, Riverside, CA 92507-4617, USA*

ABSTRACT

Predictive assessments of non-point source (NPS) pollution can have great utility for environmentally focused land use decisions related to both the remediation of existing groundwater contamination and the regulation of current (and future) agrochemical use. At the regional scales associated with NPS agrochemical applications there are staggering data management problems in assessing potential groundwater vulnerability. Geographical information system (GIS) technology is a timely tool that greatly facilitates the organized characterization of regional-scale variability. In this paper we review the recently reported (Loague *et al.*, 1998a,b) simulations of NPS groundwater vulnerability, resulting from historical applications of the agrochemical DBCP (1,2-dibromo-3-chloropropane), for east-central Fresno County (California). The Fresno case study helps to illustrate the data requirements associated with process-based three-dimensional simulations of coupled fluid flow and solute transport in the unsaturated/saturated subsurface at a regional scale. The strengths and weaknesses of using GIS in regional-scale vulnerability assessments, such as the Fresno case study, and the critical problem of estimating the uncertainties in these assessments (owing to both data and model errors) are discussed. A regional GIS-driven integrated assessment approach is proposed, which is based upon cost–benefit analysis, and incorporates both physical and economic factors that can be used in a regulatory decision process.

KEY WORDS   non-point source pollution; groundwater contamination; regulation of agrochemicals; Fresno County, California; GIS; DBCP

INTRODUCTION

Non-point source groundwater vulnerability

It is now well know that groundwater contamination is one of the US's most important environmental quality concerns. The assessment and remediation of non-point source (NPS) groundwater contamination, from the past, present and future use of agrochemicals, can easily pose problems that have significantly greater economic effects than those that have long been recognized for point sources (Loague *et al.*, 1996). In many instances, agrochemicals applied at or near the surface, with little perceived threat of contaminating the groundwater below, have leached to considerable depths. One of the greatest challenges today is to quantitatively assess the vulnerability of precious groundwater resources at regional scales, as they are affected by the long-term applications of agrochemicals that cover thousands of hectares (Kellogg *et al.*, 1992; National Research Council, 1993; Corwin and Loague, 1996; Corwin *et al.*, 1997).

---

* Correspondence to: Keith Loague, Department of Geological and Environmental Science, Stanford University, Stanford, CA 94305-2115, USA

Modern agriculture is at the root of the NPS groundwater contamination problem, i.e. large-scale agriculture depends upon agrochemicals. The goal of sustainable agriculture is to meet the needs of the present without compromising the ability to meet the needs of the future. Ideally, it strives to optimize food production while maintaining economic stability, minimizing the use of finite natural resources and minimizing environmental effects. This presents a formidable dilemma because agriculture remains as the single greatest contributor of NPS pollutants to soil and water resources (Humenik *et al.*, 1987). Point source and NPS pollutants differ in the scale of the areal extent of their source. Point source pollutants are those isolated to a single point location, such as a hazardous waste spill or a dump site. In contrast, NPS pollutants are spread across broad areas encompassing hundreds, thousands or even millions of hectares of soil and millions of litres of water. Obviously, land use decisions related to NPS groundwater contamination are very different from those for point sources. For example, it is just not possible to dig up California's San Joaquin Valley and place the contaminated soils in a secure landfill as might be the remedy for soils tainted by a leaking storage tank.

Khan *et al.* (1986) were among the pioneers in the development of regional-scale rating maps for pesticide leaching. In general, most of the NPS vulnerability assessments undertaken up until now have not considered the dynamics of a stressed saturated subsurface (i.e. groundwater pumping) at the regional scale. More importantly, leaching indices of the type used by Khan *et al.* (1986), which screen/rank chemical vulnerability, do not provide an estimate of the chemical concentrations loaded to the water table, which are essential in the decision management arena. The use of process-based numerical simulation models for the assessment of groundwater vulnerability at regional scales, albeit data and computationally intensive, is perhaps the most effective means of addressing problems that in all likelihood will not be investigated thoroughly through field study because of the lag times associated with the unsaturated zone, which can easily span several decades. For example, simulation is perhaps the easiest way to assess the lingering effect of legacies, i.e. agrochemicals long since out of use whose fate and transport are still of great concern. For the agrochemicals currently in use, and those that may be developed for use in the future, simulation allows one to look well into the future and consider alternative management strategies. The simulation effort reviewed in this paper was conducted (Loague *et al.*, 1998a,b) to estimate the regional-scale fate and transport history of DBCP (1,2-dibromo-3-chloropropane) in east-central Fresno County (California). The Fresno DBCP case study is a process-based NPS vulnerability assessment, coupling the unsaturated near-surface to the saturated subsurface, i.e. going well beyond the typical screening model approach.

Geographical information systems (GIS)

A geographical information system (GIS) is an integrating information technology that can include aspects of surface culture, demographics, economics, geography, surveying, mapping, cartography, photogrammetry, remote sensing, landscape architecture and computer science. GIS technology links the characteristics of a place, a resource and/or a feature with its spatial location. The principles and nuances of GIS techniques are reviewed by Burrough (1986). Goodchild (1993) defines a GIS as a 'general-purpose technology for handling geographic data in digital form with the following capabilities: (i) the ability to preprocess data from large stores into a form suitable for analysis (reformatting, change of projection, resampling, and generalization), (ii) direct support for analysis and modeling, and (iii) postprocessing of results (reformatting, tabulation, report generation, and mapping)'. In the context of NPS vulnerability assessments, a GIS is a tool used to characterize the full information content of the spatially variable data required by solute transport models. The advantages of GIS in its application to general spatial problems, as identified by Walsh (1988), include 'the ease of data retrieval; ability to discover and display information gained by testing interactions between phenomena; ability to synthesize large amounts of data for spatial examination; ability to make scale and projection changes, remove distortions, and perform coordinate rotation and translation; and the capability to discover and display spatial relationships through the application of empirical and statistical models'.

The use of GIS in environmental modelling has proliferated over the past two decades. However, to date, no generalized GIS system has the data representation flexibility for space and time together with the algorithmic capability needed to construct process-based models internally; consequently, environmental models and GIS must be coupled. Three of the most common strategies for linking a GIS to an environmental model (illustrated in Figure 1) are: (i) loose coupling; (ii) tight coupling; and (iii) an embedded system approach. A loose coupling involves data transfer from one to another by storage of data in one system and subsequent reading of the data by the other. The important characteristic of loose coupling is the separate functionality of the programs that implement the GIS and those that implement the models. Characteristically, a tight coupling provides a common user interface for both the GIS and the model i.e. the information sharing between the respective components is transparent. A tightly coupled model and the GIS must share the same database. As the degree of coupling between the GIS and the model increases, to the point where the model's functions are essentially part of the built-in functionality of the GIS, the model becomes embedded. In embedded systems, the coupling of software components occurs within a single application with shared memory rather than sharing the database and a common interface. The majority of the GIS linked environmental modelling applications that have been described in the literature are based upon loose coupling. The regional-scale, process-based NPS groundwater vulnerability assessment reviewed here is an example of a loosely coupled system.

AN ASSESSMENT OF REGIONAL-SCALE NON-POINT SOURCE GROUNDWATER VULNERABILITY: THE FRESNO CASE STUDY

Between the late 1950s and the time of its statewide cancellation in August 1977, there was widespread use of DBCP (1,2-dibromo-3-chloropropane) throughout the San Joaquin Valley (SJV). Almost 20 years after the cancellation, DBCP-contaminated groundwater persists as a problem in California: note, the maximum contaminant level (MCL) set for DBCP in the US for groundwater is 0.2 µg/l, the current detection limit for DBCP is approximately 0.001 µg/l.

Recently, environmental quality-driven lawsuits, for several hundred millions of dollars each (see Curtis and Profeta, 1993), have been brought against the manufacturers of DBCP related to the groundwater contamination problems in the SJV. A central issue in these cases has been whether the manufacturers of DBCP should have been expected to know that the chemical could potentially leach to groundwater and, thereby, be financially responsible for the remediation costs. The original objective of the case study (Loague et al., 1998a,b) reviewed in this paper was to address, from a simulation perspective, if 'label recommended'

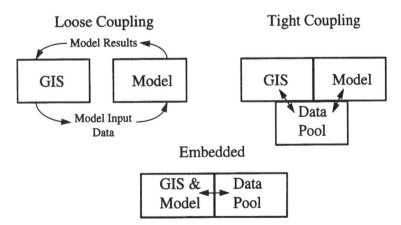

Figure 1. Three of the most common types of coupling of a GIS to an environmental model; loose, tight and embedded (after Corwin et al., 1997)

NPS applications are likely to be the principal source of the DBCP groundwater contamination in Fresno County. The Fresno case study, as presented here, is a brief summary of a two-part paper by Loague and co-workers; readers interested in the details of the modelling effort are directed to Loague *et al.* (1998a,b).

The Fresno case study boundary value problem

The San Joaquin Valley, at the southern end of California's Central Valley, extends in a south-easterly direction for approximately 400 km from just south of Sacramento. East-central Fresno County (Figure 2), situated between the San Joaquin River to the north and the Kings River to the south, is the largest agricultural county in the valley. The Fresno area is a structural depression filled with thousands of metres of sedimentary material. The estimated spatial distribution of soils in the study area is shown in Figure 3. The spatial distribution of DBCP use in the study area, between 1960 and 1977, was estimated using land cover maps for different years (e.g. Figure 4). The location of an agricultural crop on a land use map was used as an indication of the DBCP application rate for that site at that time. The DBCP application rates used for the different crops considered for the study area are given in Loague *et al.* (1998a). The water table maps constructed for the study area (e.g. Figure 5) were based upon annual estimates of the water table depth made by the Fresno Irrigation District. The areas within the study area not covered by the District's estimates were estimated for this study using nearest-neighbour extrapolation. The three-dimensional characterization of the regional geology for the Fresno study area was represented, as shown in Figure 6, by three different formations: (i) younger sediments; (ii) older sediments; and (iii) bedrock. The estimates of the formation geology and pump test information were used to characterize the distribution of saturated hydraulic conductivity (Loague *et al.*, 1998b). The pumping histories from 408 wells within the study area were included in the simulations.

The numerical model used for the one-dimensional simulations (Loague *et al.*, 1998a) of dissolved phase DBCP concentration profiles in the unsaturated zone was PRZM-2 (Mullins *et al.*, 1993). The potential fate and transport of DBCP between the surface and the water table for multiple NPS applications, related to different and changing land use between 1960 and 1977, was quantitatively estimated for 1172 elements for a 35-year period. The aggregate of the DBCP concentrations loaded to the water table for each grid element make up the annual loading files for the three-dimensional saturated transient transport simulations. The numerical models used for the three-dimensional simulations (Loague *et al.*, 1998b) of saturated subsurface

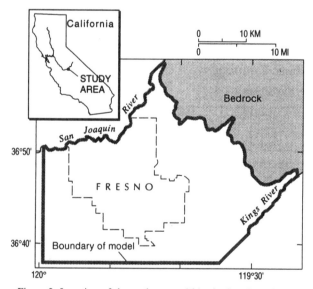

Figure 2. Location of the study area within the San Joaquin Valley

fluid flow and DBCP transport are MODFLOW (McDonald and Harbaugh, 1988) and MT3D (Zheng, 1992), respectively. Recharge to the water table for the saturated simulations was estimated as the residual (precipitation plus irrigation minus evapotranspiration) in the PRZM-2 water balance simulations. The area focused upon for the saturated simulations was represented by a three-dimensional finite-difference grid made up of 76 440 elements, i.e. 2184 1 km^2 surface elements with 35 layers. The total volume of the boundary value problem is approximately 3604 km^3.

The steps used by Loague *et al.* (1998a,b) to simulate the effect of multiple DBCP applications under changing landuse and groundwater pumping/recharge in the Fresno study area are summarized below.

1. Approximate the climatic history (1950–1994) using rainfall, temperature and pan evaporation data from the Fresno area.

2. Approximate the distribution of the major soil orders, based upon soil taxonomy and the mapped soil series.

3. Approximate the land cover history (1958-1994), based upon aerial photographs.

4. Approximate the average irrigation history (1960–1994).

5. Approximate the water table depth history (1960–1994).

6. Simulate (with PRZM-2) the transient vertical unsaturated fluid flow and DBCP transport for 1172 separate 1 km^2 elements (1960–1994).

7. Abstract the DBCP concentration at the water table for each element (1960–1994), based upon the simulated last day of the year concentration profiles.

8. Approximate the geology, based upon well logs.

9. Approximate the distribution of saturated hydraulic conductivity.

10. Approximate the recharge history (1960–1994).

11. Approximate the pumping history (1960–1994).

12. Simulate (with MODFLOW) three-dimensional groundwater flow for 35 separate years (1960–1994) i.e. one steady-state simulation for each year.

13. Simulate (with MT3D) three-dimensional saturated transient DBCP transport for 35 years (1960–1994).

The Fresno case study results

The aggregate of DBCP loaded at the water table for the entire simulation period is shown in Figure 7; Loague *et al.* (1998a) show snapshots of the water table loading for several individual years. The near-surface simulation results for the 35-year period lead to the following general comments related to the concentrations of DBCP loaded to the water table (Loague *et al.*, 1998a).

1. The areas most likely to facilitate DBCP leaching through the entire unsaturated soil profile are targeted.

2. The first appearance of DBCP above the detectable limit at the water table is simulated as most likely occurring between 1961 and 1965.

3. The estimated concentrations reaching the saturated subsurface exceed the MCL at several locations at different times. The first appearance above the MCL was between 1965 and 1970; by 1990, the concentrations are below the MCL.

4. Relative to the size of the study area, the extent and duration of the estimated contamination is small.

5. Concentrations are a function of the application rates and frequency; the highest estimates are associated with citrus and vineyards, the lowest estimates are associated with cotton.

6. Concentrations are a function of the unsaturated profile thickness. The longer the residence time the chemical has in the unsaturated zone the more opportunity there is for decay and vapour diffusion.

7. Concentrations are a function of soil hydraulic properties. The higher estimates are associated with the Alfisols and Entisols, the lower estimates are associated with the Inceptisols, Mollisols and Vertisols.

8. Concentrations are a function of near-surface sorption. The higher estimates are associated with the Entisols and Vertisols, the lower estimates are associated with Mollisols.

Figure 8 illustrates the simulated water table elevations for the study area for a single year and the simulated concentrations for the DBCP plume for three individual years; Loague et al. (1998b) show observed versus predicted water table elevations and DBCP concentrations for several individual years. The simulation results for the saturated subsurface for the 35-year period lead to the following general comments related to the fate and transport of DBCP in the study area (Loague et al., 1998b).

1. The MODFLOW and MT3D simulations appear, in a qualitative sense, to be quite reasonable without the aid of calibration.

2. The direction of groundwater flow and the movement of the plume is from the east to west.

3. The plume evolves (grows and retracts) with time owing to the loading rates at the water table.

4. The plume does not migrate into the older sediments during the simulation period.

5. The simulated concentrations, for the base case, are below the MCL.

6. The shape and movement of the plume correlates well with the estimated application history.

DISCUSSION

The Fresno case study and GIS

The coupled modelling approach employed for the Fresno case study facilitated the simulation of an extremely complicated problem in an efficient manner. GIS facilitated the preparation of the individual data overlays (e.g. soils, land cover, water table depth, geology). In addition, GIS greatly enhanced the presentation of the simulation effort reported here and by Loague et al. (1998a,b). For example, Figures 3–6 begin to show the variability of the tremendous amount of information needed to excite the models, while Figure 7 shows the variability in the near-surface leaching results.

The PRZM-2 simulations were designed to characterize the non-intuitive interplay between climate, soil type, land use and chemical properties as related to the long-term leaching of DBCP in the Fresno study area. The results from the near-surface portion of the Fresno case study (Loague et al., 1998a) illustrate, for the best estimates of historical chemical use patterns, soil properties and water table depths, that DBCP can be expected to leach to groundwater in Fresno County as a result of regional-scale, agricultural-related applications. The simulated leaching history for DBCP in Fresno County suggests that the pesticide has migrated from the surface to the water table in detectable concentrations and at some locations these concentrations exceed the MCL. The results suggest that future DBCP loading at the water table, resulting from past applications, will be below the detectable limit.

The MODFLOW and MT3D simulations were designed to characterize the non-intuitive interplay between DBCP loading at the water table, recharge, pumping and regional groundwater flow as related to the long-term fate and transport of DBCP in the Fresno study area. The results from the saturated sub-surface portion of the Fresno case study (Loague et al., 1998b) illustrate the three-dimensional evolution of the DBCP plume in the Fresno study area over a 35-year period based upon our best estimates of hydro-geological parameters, water table depths and pumping rates. The simulated DBCP concentrations are limited to the relatively shallow younger sediments and are generally well below the MCL. The simulations suggest that NPS applications of DBCP are not responsible for the observed 'hot spots' in the study area (see

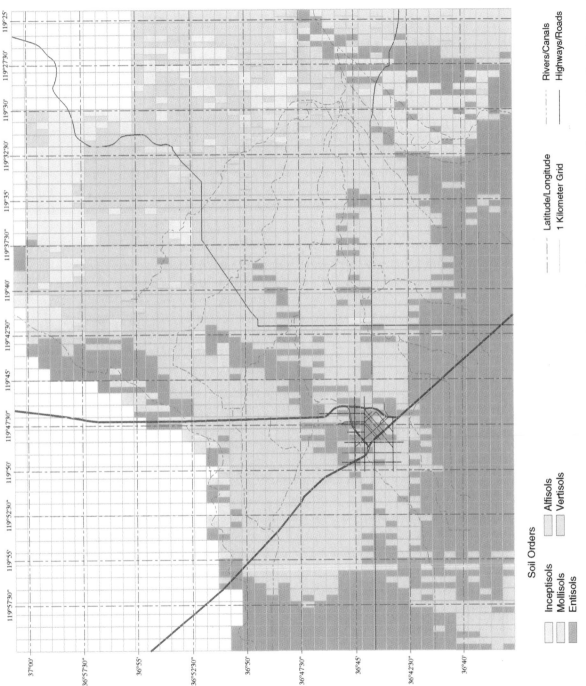

Figure 3. The distribution of soil orders within the study area. The study area corresponds, approximately, to east-central Fresno County; the location of the city of Fresno is earmarked by the road intersections in the lower left quarter of the map

Soil Orders

Inceptisols   Alfisols
Mollisols   Vertisols
Entisols

Latitude/Longitude   Rivers/Canals
1 Kilometer Grid   Highways/Roads

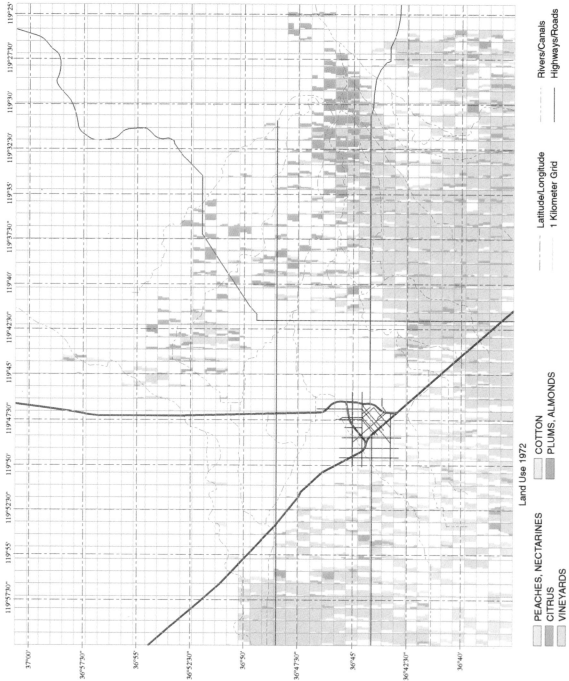

Land Use 1972

PEACHES, NECTARINES   COTTON   Latitude/Longitude   Rivers/Canals

CITRUS   PLUMS, ALMONDS   1 Kilometer Grid   Highways/Roads

VINEYARDS

Figure 4. Land cover representation (1972) for the study area

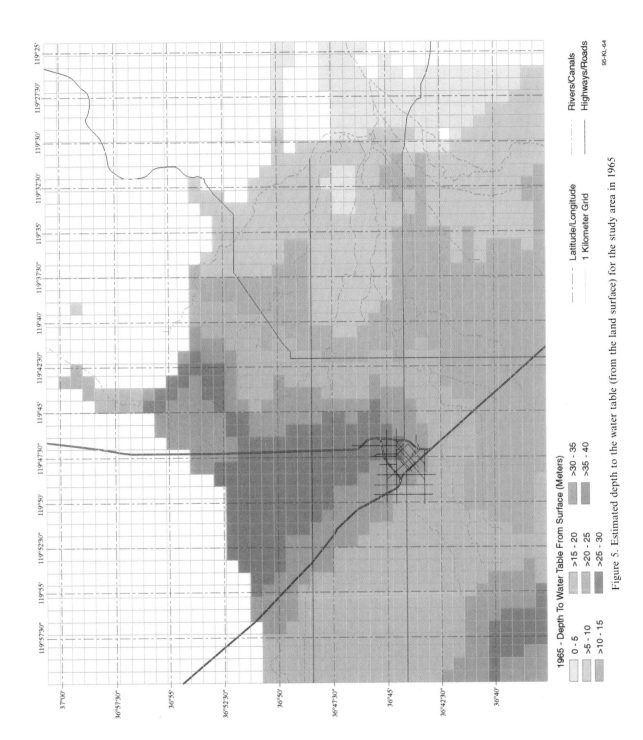

1965 - Depth To Water Table From Surface (Meters)

0 - 5	>30 - 35
>5 - 10	>35 - 40
>10 - 15	
>15 - 20	------ Latitude/Longitude
>20 - 25	—— 1 Kilometer Grid
>25 - 30	

------ Rivers/Canals

—— Highways/Roads

95-KL-64

Figure 5. Estimated depth to the water table (from the land surface) for the study area in 1965

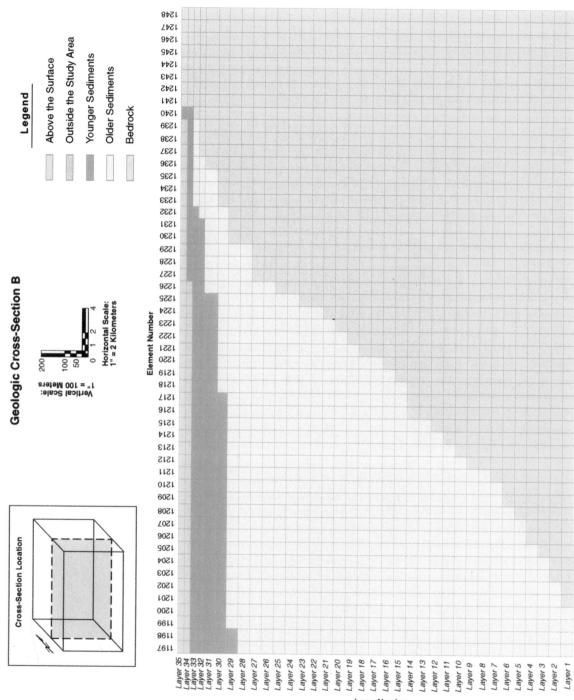

Figure 6. Estimated east–west cross-section of the study area geology. The upper portion of the sedimentary sequence is dominated by late Cenozoic stream deposits which contain the water-bearing zones of interest in the simulation effort (see Loague *et al.*, 1998b)

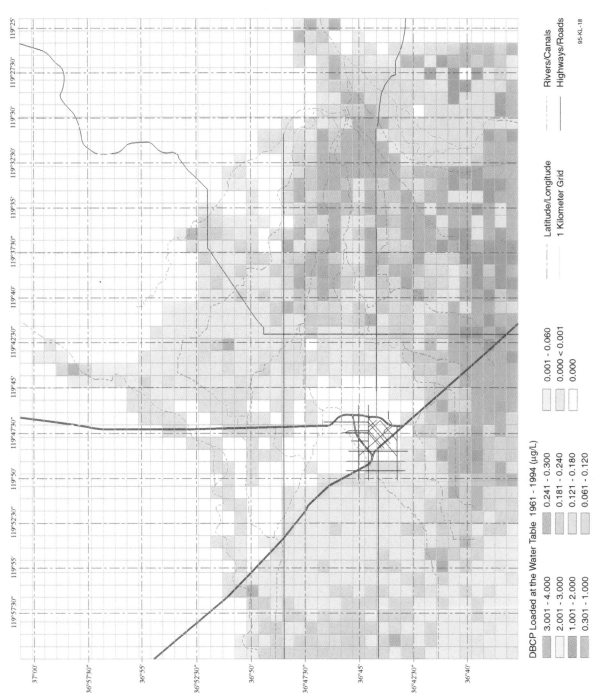

DBCP Loaded at the Water Table 1961 - 1994 (µg/L)

	3.001 - 4.000		0.241 - 0.300		0.001 - 0.060
	2.001 - 3.000		0.181 - 0.240		0.000 < 0.001
	1.001 - 2.000		0.121 - 0.180		0.000
	0.301 - 1.000		0.061 - 0.120		

Latitude/Longitude

1 Kilometer Grid

Rivers/Canals

Highways/Roads

95-KL-18

Figure 7. Total simulated DBCP loading at the water table between 1960 and 1994 for the study area

(a)

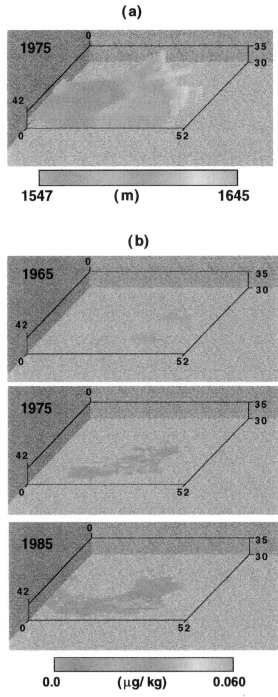

Figure 8. (a) Simulated water table elevations for the study area for 1975. (b) Simulated DBCP concentrations for the study area for 1965, 1975 and 1985

Loague *et al.*, 1998b); it is quite possible that these isolated high concentrations are a result of point source loading (e.g. spills at field mixing sites, down-well dumping and/or burial).

Uncertainties in regional-scale assessments

Why model? In general, there are two idealized uses for simulation in hydrology. The first is the prediction (or forecasting) of future events based upon a calibrated and validated model. The second use is the development of concepts for the design of future experiments to improve the understanding of processes. The Fresno case study reviewed in this paper was intended for both prediction and concept development.

There are three sources of error inherent to modelling in hydrology: (i) model error; (ii) input error; and (iii) parameter error. Model error results in the inability of a model to simulate the given process, even with the correct input and parameter estimates. Input error is the result of errors in the source terms (e.g. soil water recharge and chemical application rates). Input error can arise from measurement, juxtaposition and/ or synchronization errors. Parameter error has two possible connotations. For models requiring calibration, parameter error is usually the result of model parameters that are highly interdependent and non-unique. For models with physically based parameters, parameter error results from an inability to represent aerial distributions on the basis of a limited number of point measurements. The aggregation of model error, input error and parameter error is the total (or simulation) error. For multiple-process and comprehensive models, simulation error is complicated further by the propagation of error between model components.

In general, the methods for characterizing uncertainty can be grouped into three categories (Loague and Corwin, 1996): (i) first-order analysis; (ii) sensitivity analysis; and (iii) Monte Carlo analysis. First-order analysis is a simple technique for quantifying the propagation of uncertainty from input parameter to the model output. The first-order approximation of functionally related variables is obtained by truncating a Taylor series expansion (about the mean) for the function after the first two terms. Sensitivity analysis is used to measure the effect that changing one factor has on another. The sensitivity of a model's output to a given input parameter is the partial derivative of the dependent variable with respect to the parameter. Monte Carlo analysis is a stochastic technique for characterizing the uncertainty in complex hydrological response model simulations. The Monte Carlo method considers each model input parameter to be a random variable with a probability density function (PDF). Monte Carlo simulations are based upon a large number of realizations, from every input parameter distribution, created through sampling the different PDFs with a random number generator. A separate hydrological response simulation is made for each parameter realization. The number of possible simulations, based upon all the combinations of parameter realizations, is infinite; therefore, a finite number of cases (usually several hundred) are usually investigated. Estimates of the average simulated hydrological response, and the associated uncertainty, are made from the combined outputs of the simulations (i.e. the total ensemble of the different realizations). Loague and Corwin (1996) provide examples of first-order uncertainty analysis, sensitivity analysis and Monte Carlo simulation.

Assessments of NPS groundwater vulnerability rest upon soil, climatic and chemical data that are extremely sparse at regional scales and, therefore, contain considerable uncertainty. The implications of the uncertainties associated with data and model errors, as well as data worth consideration, are discussed in the review by Loague *et al.* (1996) for regional-scale groundwater vulnerability assessments in Hawaii. The ongoing phase in the coupled simulation effort discussed in this paper is the quantitative characterization of the uncertainties (i.e. model and data errors) for the regional-scale NPS assessment of groundwater vulnerability for east-central Fresno County.

Risk assessment for regional-scale vulnerability

Management decisions that result from environmental policy assessments of NPS pollution require an approach that integrates physical and social sciences in a decision framework. Conceptual economic models for estimating the effects of agricultural production and environmental residuals have been developed, in a point source framework, in parallel with, but not rigorously coupled to, physics- and chemistry-based process models of near-surface solute transport. These economic models are based on an individual farm unit's production. The farm unit models are used to evaluate agricultural and/or environmental policy issues both on an individual basis and on a regional scale. While these models are of value, we do not consider them to be relevant for regional-scale policy analysis because of the uncertainties inherent in spatial environmental data.

At a regional scale, the utility of scientific data lies in the ability to screen areas of greater (lesser) potential for groundwater contamination. Instead of collecting soils information at a sufficient number of sites to be able to apply geostatistics efficiently for identification of spatial structure, for use in GIS-based regional-scale environmental risk assessment, Loague *et al.* (1996) propose a regional integrated assessment (RIA) approach. The RIA method, based on cost–benefit analysis, incorporates both physical (spatial point processes) and economic (states of nature) variables in a regional screen that can be included as a protocol in a regulatory decision process. The screening process uses GIS as an analytical tool to identify areas that have a relatively high probability for loss based upon a variety of physical factors. For example, physical factors, such as mobility indices, may be used to represent dynamic hydrological processes. The model is then cast in a decision framework and follows a generic procedure. The results of an RIA would be presented as an environmental risk map. The following four steps make up an RIA evaluation (Loague *et al.*, 1996).

1. Areas that would require mitigation are identified.

2. For each area, a probabilistic hazard map (includes mean and variance) is estimated and compiled as a function of physical attributes.

3. The cost of the required mitigation in each area is estimated.

4. For each area identified as requiring mitigation, the expected mitigation cost (loss) avoided is estimated as the product of the probability of the hazard and the cost of mitigation.

CONCLUSIONS

GIS-based models of NPS pollutants in subsurface soil and water systems have proliferated over the past decade. The acknowledged trend in the coupling of GIS with environmental models will continue at an even greater pace because of the introduction of cheaper, desktop GIS software that is customizable to the application and to the growing demand for spatial environmental information. Without question, GIS is serving as a catalyst to bring solute transport modelling, data acquisition and spatial databases into a self-contained package to assess NPS pollutant problems. Yet, a cautionary footnote is needed, because the sophisticated visualizations created from GIS should never disguise the legitimacy of the rendered results nor should simulated results ever supplant field observation (Corwin *et al.*, 1998).

The case history simulations discussed in this paper provide a good starting point for understanding the regional-scale fate and transport of DBCP in east-central Fresno County, which could not have been obtained without the coupled process-based simulation effort and GIS. It is our opinion that for regional-scale groundwater vulnerability assessments to be of any significance relative to future regulation of agrochemical use the approaches to be developed must include: (i) concentration profile estimates in time and space; (ii) epidemiologically based contamination levels of concern relative to human health; and (iii) economic constraints. The next step in our own work is to incorporate economic constraints into stochastic–conceptual groundwater vulnerability assessments at regional scales.

ACKNOWLEDGEMENTS

This paper is a CESIR contribution. We are grateful to Robert Abrams, Stanley Davis, D'Artagnan Lloyd, Anh Nguyen, Laura Serna, Iris Stewart and Erik Wahlstrom for their many efforts related to the Fresno case study.

REFERENCES

Burrough, P. A. 1986. *Principles of Geographic Information Systems for Land Resources Assessment*. Oxford University Press, Oxford. 193 pp.

Corwin, D. L. and Loague, K. (eds) 1996. *Applications of GIS to the Modeling of Non-Point Source Pollutants in the Vadose Zone*, Special Publication 48. Soil Science Society of America, Madison. 319 pp.

Corwin, D. L., Vaughan, P. J., and Loague, K. 1997. 'Modeling nonpoint source pollutants in the vadose zone with GIS', *Environ. Sci. Technol.*, **31**, 2157–2175.

Corwin, D. L., Loague, K., and Ellsworth, T. R. 1998. 'GIS-based modeling of nonpoint source pollutants in the vadose zone', *J. Soil Water Conserv.*, **53**, in press.

Curtis, J. and Profeta, T. 1993. *After Silent Spring: The Unsolved Problems of Pesticide Use in the United States*. National Resources Defense Council, New York. 56 pp.

Goodchild, M. F. 1993. 'The state of GIS for environmental problem-solving', in Goodchild, M. F., Parks, B. O., and Steyaert, L. T. (eds), *Environmental Modeling with GIS*. Oxford University Press, New York. pp. 8–15.

Humenik, F. J., Smolen, M. D., and Dressing, S. A. 1987. 'Pollution from nonpoint sources: where we are and where we should go', *Environ. Sci. Technol.*, **21**, 737–742.

Kellogg, R. L., Maizel, M. S., and Goss, D. W. 1992. *Agricultural Chemical Use and Ground Water Quality: Where are the Potential Problem Areas*. National Center for Resource Innovations, USDA, Washington, DC. 41 pp. plus appendices.

Khan, M. A., Liang, T., Rao, P. S. C., and Green, R. E. 1986. 'Use of an interactive computer graphics and mapping system to assess the potential for groundwater contamination with pesticides', *EOS, Trans. Am. Geophys. Union*, **67**, 278.

Loague, K. and Corwin, D. L. 1996. 'Uncertainty in regional-scale assessments of non-point source pollutants', in Corwin, D. L. and Loague, K. (eds), *Applications of GIS to the Modeling of Non-Point Source Pollutants in the Vadose Zone*, Special Publication 48. Soil Science Society of America, Madison. pp. 131–152.

Loague, K., Bernknopf, R. L., Green, R. E., and Giambelluca, T. W. 1996. 'Uncertainty of groundwater vulnerability assessments for agricultural regions in Hawaii: review', *J. Environ. Qual.*, **25**, 475–490.

Loague, K., Lloyd, D., Nguyen, A., Davis, S. N., and Abrams, R. H. 1998a. 'A case study simulation of DBCP groundwater contamination in Fresno County, California: 1. Leaching through the unsaturated subsurface', *J. Contam. Hydrol.*, **29**, 109–136.

Loague, K., Abrams, R. H., Davis, S. N., Nguyen, A., and Stewart, I. T. 1998b. 'A case study simulation of DBCP groundwater contamination in Fresno County, California: 2. Transport in the saturated subsurface', *J. Contam. Hydrol.*, **29**, 137–163.

McDonald, M. G. and Harbaugh, A. W. 1988. *A Modular Three-Dimensional Finite-Difference Groundwater Flow Model*. Scientific Software Group, Washington, DC.

Mullins, J. A., Carsel, R. F., Scarbough, J. E., and Ivery, A. M. 1993. *PRZM-2, A Model for Predicting Pesticide Fate in the Crop Root and Unsaturated Soil Zones: Users Manual for Release 2*, EPA-600/R-93/046. Environmental Research Laboratory, USEPA, Athens.

National Research Council, 1993. *Ground Water Vulnerability Assessment: Predicting Relative Contamination Potential under Conditions of Uncertainty*. Nation Academy Press, Washington, DC. 204 pp.

Soil Survey Staff, 1975. *Soil Taxonomy*, USDA-SCS Agric. Handbook No. 436. US Government Printing Office, Washington, DC.

Walsh, S. J. 1988. 'Geographic information systems: an instrumental tool for earth science education', *J. Geog.*, **87**, 17–25.

Zheng, C. 1992. *MT3D: A Modular Three-Dimensional Model for Simulation of Advection, Dispersion, and Chemical Reactions of Contaminants in Groundwater Systems*, Version 1·5. S. S. Papadopulos & Assoc., Bethesda, MD.

11

SYNOPTIC VIEWS OF SEDIMENT PLUMES AND COASTAL GEOGRAPHY OF THE SANTA BARBARA CHANNEL, CALIFORNIA

LEAN A. K. MERTES,* MELODEE HICKMAN, BEN WALTENBERGER, AMY L. BORTMAN, ETHAN INLANDER, CERETHA McKENZIE AND JOHN DVORSKY

Department of Geography and Institute for Computational Earth System Science, University of California, Santa Barbara, CA 93106-4060, USA

ABSTRACT

Representative patterns of surface sediment plumes in coastal waters of the Santa Barbara Channel are described based on analysis of a time-series of remote sensing data (1972–present) for winter conditions. The first-order terrestrial influence on the patterns is that the largest plumes are generated from the largest watersheds during and following significant (> 3 cm precipitation) winter storms. For example, coastal sediment plumes generated from the three largest watersheds, the Santa Ynez, Ventura and Santa Clara river watersheds, are always the largest surface plume features in the winter coastal waters. The largest plume observed in the remote sensing data was generated from the combined outflow of the Ventura and Santa Clara rivers in February 1983, following approximately 25 cm of rain in the previous month, and averaged 10 km long and 25 m wide. A second-order effect on the spatial patterns of the coastal plumes is the position of watershed outlets with respect to coastal geography as shown by a comparison of, (i) the watershed morphometry, (ii) the accumulated runoff, (iii) the cell-based erosion potential, and (iv) the near-shore bathymetry for two of the Northern Channel Islands, Santa Rosa and Santa Cruz. In particular, Santa Rosa Island is surrounded by shallow coastal waters which may experience resuspension of sediment throughout the year.

KEY WORDS   coastal processes; watershed analysis; GIS hydrology; remote sensing

INTRODUCTION

Resource management of coastal environments requires an interdisciplinary consideration of both marine and terrestrial influences (Holligan and Reiners, 1992) in order to take account of the entire range of processes influencing the functioning of the coastal area. Examination and management of processes within this ecosystem presents a set of scientific challenges that can be partially met through combining field data, modelling and digital technologies. Digital databases, remote sensing data and spatial analysis tools embedded in a geographical information system (GIS) provide opportunities to archive spatially referenced data that can be analysed for relationships between environmental variables. For example, remote sensing data provide a synoptic view of large areas at coastal margins, while GIS software provides a framework for implementing the stratification techniques of watershed analysis required to quantify and interpret the land use, hydrology and geomorphology of coastal watersheds. The results of this type of analysis are described here and are the first step towards quantifying patterns in these coastal waters based on interpretation of terrestrial processes, such as storm patterns and their effect on sediment transport in coastal watersheds.

* Correspondence to: Leal A. K. Mertes, Department of Geography and Institute for Computational Earth System Science, University of California, Santa Barbara, CA 93106–4060, USA. Email address: L. A. K. Mertes (leal@geog.ucsb.edu).

Contract grant sponsors: NOAA, UCSB, Office of Naval Research.

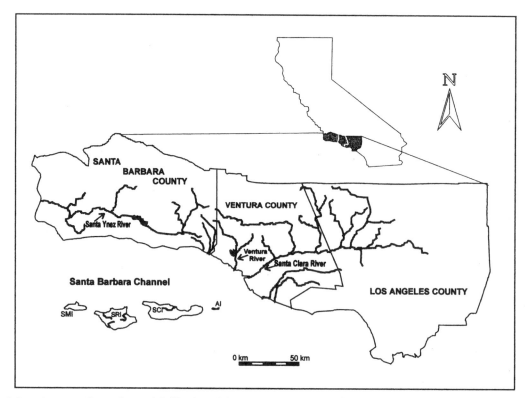

Figure 1. Location map of central coastal California and the Northern Channel Islands. Map shows major watersheds contributing to the Santa Barbara Channel

In particular, this study investigates the types and relative magnitude of processes responsible for the transfer of sediment from ridges to the ocean for several coastal river systems (Figure 1) of the semi-arid, tectonically active terrain of the Transverse Ranges of California. In these coastal watersheds, a flashy, often intense, precipitation regime (Figure 2), along with rapidly rising, rugged topography, fires and fractured sedimentary rocks combine to yield extremely large sediment loads to the ocean (Scott and Williams, 1978; Milliman and Syvitski, 1992). Owing to the seasonal fluctuations in precipitation of this mediterranean climate, sediment transfer is a pulsed phenomenon typically limited to the months December to February (Figure 2), with storm activity apparently exacerbated by El Niño conditions (Haston and Michaelsen, 1994). In addition, sediment from several of these watersheds is transferred into the anoxic waters of the Santa Barbara Basin, where 100 000 years of well-preserved varves record seasonal and longer term pulsed characteristics (Schimmelmann *et al.*, 1990; Kennedy and Brassell, 1992; Behl and Kennett, 1996).

Using the best available remote sensing data since 1972 (the onset of Landsat image data acquisition) we assembled a time-series of images that represent a range of winter storm conditions for the Santa Barbara Basin. Analysis of these remote sensing data shows, not unexpectedly, that the first-order effects on the patterns of sediment plumes in this coastal area are storm size and the relative size of the locally contributing watershed. Second-order effects on differences in the plumes may be related to basin morphometry and the spatial correlation of watershed outlets and coastal geography. To examine this hypothesis, a GIS-based analysis of (i) watershed morphometry, (ii) accumulated runoff, (iii) cell-based erosion potential, and (iv) near-shore bathymetry for two of the islands in the Santa Barbara Channel was completed. The results show that the juxtaposition of stream outlets with shallow coastal waters for one of the islands allows the development of extensive surface plumes that probably are as much related to sediment outflow as to resuspension of sediment in the coastal zone.

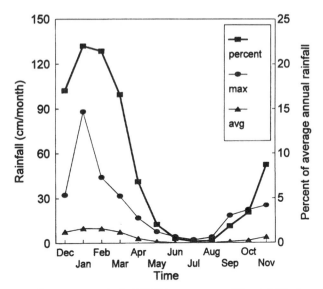

Figure 2. Average rainfall parameters for water years 1868–1995. The average monthly rainfall is shown by the filled triangles, while the maximum monthly total for the entire record is shown as filled circles. Seasonality of precipiation is demonstrated by the percentage of total rainfall shown for each month as the thick line with filled squares. Data for the Santa Barbara sewage gauge provided by D. Gibbs (Santa Barbara County Flood Control) and J. Michaelsen

STUDY AREA

Climate

The study region is characterized by a mediterranean climate with mild, moist winters and moderately warm, generally rainless summers (Figure 2). Average monthly temperatures range from 10°C (50°F) in January to 18°C (65°F) in August (Ferren *et al.*, 1990). Higher temperatures prevail in the summer months in the interior, often exceeding 38°C (100°F). The presence of coastal fog is primarily responsible for the narrow seasonal range in temperature. Rates of precipitation are highest at the highest elevations, with approximately 100 cm of precipitation at the higher elevations and 40 cm of precipitation near the coast. Floods in these watersheds result from intense storms that can be exacerbated by El Niño atmospheric conditions (Haston and Michaelson, 1994). Typically, there is a very short time lag (several hours) from rainfall to runoff owing to the steep terrain, intensity of the rainfall and thin soil cover.

Geology

The study region rises steeply to over 2500 m on the mainland and over 750 m on the Northern Channel Islands. The rugged topography of steep slopes and narrow canyons is the result of tectonic activity in the Transverse Ranges that are in an anomalous 'east–west trending topographic and structural zone superposed on the otherwise north–west-trending structural grain of California' (Rockwell *et al.*, 1984, p. 1466). The Northern Channel Islands are a structural element of the Western Transverse Ranges Province and are considered an extension of the Santa Monica Mountains to the south of the study area (Patterson, 1979; Dibblee, 1982). Average areal rates of uplift of less than 1 mm/yr to over 1 cm/yr have been reported by Scott and Williams (1978), Keller *et al.* (1982), and Rockwell *et al.* (1984). Since the Pasadenan orogeny in the middle Pleistocene, several major geological structures have continued to transform the most recent deposits and provide the structural framework controlling the watershed patterns and landscape development of the study region (Scott and Williams, 1978).

The lithology of the mainland coastal watersheds is comprised primarily of Cretaceous to Holocene sedimentary rocks that are largely tilted, overturned or otherwise deformed (Putnam, 1942; Scott and Williams, 1978). The area has an extremely thick sequence of largely marine sedimentary rocks (Ferren *et al.*, 1990), which exceeds 18 000 m (Putnam, 1942). The Santa Clara Basin has been particularly noted for its wide diversity of rock types (e.g. Ingersoll *et al.*, 1993). Alternating sequences of more resistant sandstone and less resistant shales have resulted in alternating steep and gentle terrain throughout the study area. The valley bottoms tend to be filled with alluvium except in the highest parts of the drainages where the bedrock is exposed in channels. The Northern Channel Islands are a mix of volcanic and volcaniclastic sediments of Miocene age and pre-Cretaceous crystalline basement, which are typically overlain by marine and non-marine sedimentary and volcaniclastic Tertiary deposits (Patterson, 1979; Dibblee, 1982; Jones and Grice, 1993).

Water and sediment transport

The study area includes the drainage basins of several rivers of the Transverse Ranges and Northern Channel Islands (Figure 1). The mainland areas have been studied in greater detail than the Channel Islands. However, the climate, vegetation, topography and lithology are similar enough that the same types of processes may be active in the different locations. The drainage basins range in size from small drainages flowing directly into the ocean with areas of less than 10 km^2, to the Santa Clara River drainage basin, of approximately 4100 km^2 (Fan, 1976). The two largest Northern Channel Islands, Santa Cruz and Santa Rosa islands, are 249 and 214 km^2 in size, respectively.

The following summary is based on studies completed in mainland coastal watersheds. Stream channels typically fill with sediment during drier periods, and are scoured during floods (Scott and Williams, 1978; Keller and Capelli, 1992; Keller *et al.*, 1997). The effects on rates of sediment erosion and transport include the effects of wildfires. Florsheim *et al.* (1991) and Rice (1982) concluded that chaparral wildfires with a recurrence interval of 30–65 years contribute to sediment yield and redistribution primarily by causing an increase in the rate of sediment delivery to channels by dry ravelling. They concluded that fires do not always result in large debris flows, because these types of flows require the coincidence of intense rainfall and recent burns. The recurrence intervals of large debris flows in the basin appear to be an order of magnitude greater (approximately 500 years) than for fires, based on radiocarbon dates of debris flow deposits (Best and Keller, 1986; Florsheim *et al.*, 1991). Based on field measurements of channel morphology in two watersheds affected by the 1990 Painted Cave Fire, Keller *et al.* (1997) concluded that the primary movement of sediment after a fire is caused by flood flows and not debris flows. The influence of human-induced effects, such as dams and reservoirs, timber harvest, urbanization and gravel mining in this coastal area (Kondolf and Keller, 1991) is not yet well understood.

On Santa Cruz Island, Brumbaugh (1980) interpreted several stratigraphic sections of channel deposits that show a fine alluvium capped with coarse, poorly sorted alluvium as evidence for a change in the sediment transport processes during historical time. The intensity of grazing on Santa Cruz Island has fluctuated over the past two centuries with a resulting loss in vegetation cover. 'The slopes, overgrazed and made barren by sheep, may have become dominated by rilling and mass wasting ...' (Brumbaugh, 1980, p. 149) which has produced this coarser surface alluvium. Recent reductions in the sheep population have resulted in some recovery of vegetation (Brumbaugh, 1980; L. Mertes and M. Cobb, unpublished data), but no study has been completed to determine if there has been a reduction in sediment erosion. Renwick (1982) documented several types of mass movements on Santa Cruz Island as the result of intense rainfall during the winter of 1977–1978. The factors controlling the extent and type of mass wasting included lithology, topography, soil type and vegetation cover. In particular, the permeability of soils, as interpreted from soil texture data, seemed to be the most important factor for determining the morphology and frequency of landslides.

In their study of erosion rates in the Transverse Ranges, Scott and Williams (1978) concluded that the watersheds of the Ventura River Basin have the highest rates of erosion per unit area in these mountains, because of their relative steepness and weak rocks. Taylor (1983) estimated that under natural conditions, sediment delivery of 900 000 m^3/yr yields a denudation rate across the entire basin of the order of a few millimetres per year, which is somewhat less than uplift rates (Scott and Williams, 1978; Taylor, 1983).

Although average erosion rates may be extremely high, the transfer to the ocean is mitigated by the presence of estuaries and coastal marshes at the mouths of some of the watersheds. Keller *et al.* (1997) estimated the total sediment transport for the two years following the Painted Cave Fire for watersheds affected by the fire. The total transport was estimated to be 147 000 m³ of sediment, with 108 000 m³ (73%) of the total stored in the estuary after two years. If the bulk density of the sediments is assumed to be 1·6 g/cm³, then this amounts to an average annual rate of erosion of 96 t/ha over the 1190 ha of the watershed. Cole and Liu (1994) described the Holocene conditions of deposition in a small estuary on Santa Rosa Island. The rates of deposition in this estuary, which has a drainage basin area of approximately 480 ha, ranged from 0·07 mm/yr prior to 1800 AD to 24 mm/yr during the most intensive grazing period between 1874 and 1920 AD. These deposition rates yield much lower rates of erosion, i.e. 0·06 t/ha. However, unlike Keller *et al.* (1997), Cole and Liu (1994) only reported on rates of deposition in the estuary and did not report transport measurements for the entire watershed.

Once past the coastal marshes, the sediment is transported to beaches and into the deep ocean basin. The dominance of the largest watersheds is shown by provenance studies completed for the coastal environments. Drake *et al.* (1972) reported on the distribution of total suspended solids collected throughout the basin in May 1969. Concentrations ranged from < 1 mg/l to approximately 5 mg/l near the shore around the mouths of the Santa Clara and Ventura rivers. Rice *et al.* (1976) concluded, from analysis of beach sediments collected from 1967 to 1972, that the sand composition of the beaches around the Santa Barbara Basin is dominantly controlled by the sedimentary rocks of the Transverse Ranges. Beaches in close proximity to the Santa Clara River show a dominance of granodiorite, which would have eroded from the headwaters of the basin. In the ocean sediments, Marsaglia *et al.* (1995) interpret sand layer mineralogy from Ocean Drilling Program Site 893 in the Santa Barbara Basin to show a provenance associated with the Santa Clara and Ventura river outflows. Detailed provenance studies by Fan (1976) also suggest a dominance of sediment from the Santa Clara River drainage in shelf samples of the Santa Barbara Basin. Kolpack and Drake (1984) estimated that 54 million metric tons of fine-grained sediment were transferred from the Santa Clara and Ventura rivers to the deep basin as a result of storms in 1969.

DATA SETS AND METHODS

Field data and methods

Rainfall data were acquired from Santa Barbara County Flood Control for the mainland watersheds (D. Gibbs, unpublished data). For Santa Cruz Island, short-term rainfall records (1990 to present) were provided for a rain gauge with daily readings maintained by the Santa Cruz Island Natural Reserve in the central part of the island (L. Laughrin, unpublished data).

Field collection of total suspended matter (TSM) for the Santa Clara–Ventura rivers' plume took place for three days, 26–28 January 1997. Nisken bottle samples were collected at the surface and 250 ml aliquots were filtered on tared filters with a nominal pore size of 0·06 mm. Filters were washed with distilled water, dried and weighed to obtain mass per volume concentrations according to protocols described by Mueller and Austin (1995).

Remote sensing data and methods

Remote sensing data were used to characterize the locations and relative surface concentrations of plumes in the Santa Barbara Channel and to generate a land cover map for Santa Cruz Island. The Landsat record was examined for 1972 to the present for the best, cloud-free images that followed significant winter storms. The Landsat data presented here (Figure 3) were selected as representative of a range of plume patterns and storm conditions, and include Landsat MultiSpectral Scanner (MSS) data for 4 January 1983 and 19 February 1983, and Landsat Thematic Mapper (TM) data for 9 February 1994. Advanced High Resolution Radiometer (AVHRR) data were recorded on 29 January 1997 and are included because field data were collected during the three days prior to the image acquisition and provide a calibration for the sediment concentration as described below.

For analysis of the plume concentrations a statistical technique named spectral mixture analysis (Mertes *et al.*, 1993) was applied to either radiance (Landsat) or calibrated reflectance (AVHRR; E. Fields and D. Siegel, unpublished data) data in order to estimate the extent of, and particulate concentration for, plumes in each image. The calibration of the particulate concentrations in this procedure relies on the use of radiance or reflectance data that characterize both high and low sediment concentrations. These data are named end-members and can be selected directly from an image or from a catalogue of reference data. Preliminary analysis of these coastal data shows that it is not appropriate to use the end-member data described by Mertes *et al.* (1993) and that new reference end-members need to be developed for these types of particulates, based on the optical characteristics of the particles and water. Because the reference end-member data are not yet developed, image end-members and a field calibration for the concentration range were used. Therefore, in the analysis reported here, the concentration estimates for all four images are based on the end-members from the January AVHRR image for which field data were available. Analysis of the AVHRR data included the use of channels 1, 2 and 3. Analysis of the TM data included the use of only bands 1, 2, 3 and 4, while all four of the MSS bands were used. For clarity all pixels without a significant water component are masked with black.

Land cover analysis for Santa Cruz Island was achieved through a cluster analysis in ARC/Info/Grid (AIG) of a 20 October 1993, Landsat TM image. The cluster analysis yielded seven clusters, two of which were in the ocean. Each cluster was named to a broad community type, e.g. dense woodland, based on comparisons with aerial photographs and a vegetation map provided by Minnich (1980). The 1993 land cover map has all of the 'vegetation' categories listed by Jones and Grice (1993) in their Table 2, including, grass, oaks, coastal sage, chaparral, pines and bare. Chaparral and pines were combined into a dense woodland group in the current study.

Other digital data

United States Geological Survey digital elevation (30 m spatial resolution) and digital line graph data at 1 : 24 000 and 1 : 100 000 scale formed the basis for the topographic and morphometric analysis. NOAA GEODAS (Version 3·2, NOAA, 1996) point bathymetric soundings were used to generate the bathymetric surface. The raw point data are predicted to have a *z*-error (vertical) ranging from 0·3 to 0·8 m in depth, which increases with depth. The bathymetric point data were transformed to a triangulated irregular network (TIN) using the Delauney transformation at a 90 m spatial resolution. Geological data for Santa Rosa and Santa Cruz islands were digitized from 1 : 250 000 scale maps (Weaver *et al.*, 1969). Digitized vegetation maps of Santa Rosa Island (K. Schwemm, unpublished data) were based on maps produced by Clark *et al.* (1990).

GIS methods

ARC Macro Language programs (AMLs) in AIG were written to complete a morphometric analysis of the watershed study areas, to interpolate a rainfall surface based on elevation, to characterize subwatersheds, to estimate storm event runoff and to calculate potential cell-based sediment erosion. The morphometric analysis is based on cell-based modelling techniques described for AIG. These methods are elucidated by Maidment (1995). Without the data to complete a detailed water balance, we chose to convert rainfall to runoff based on standard Soil Conservation Service methods for determining hydrological response units (both at subwatershed and grid-cell scales) as outlined by Dunne and Leopold (1978). The runoff model requires land cover and soil inputs that allow for generalized estimates of soil hydrology (for categories from high to low; e.g. see Dunne and Leopold, 1978, Table 10–4) and effectiveness of the land cover in retaining rainfall (multiple categories based on generalized land cover characteristics, e.g. see Dunne and Leopold, 1978, Tables 10–3 and 10–5). The user assigns codes to the land cover and soil types. The soil and cover grids were then summed using map algebra to produce a soil cover grid. Each cell in the soil cover grid is then assigned to the appropriate runoff curves (Dunne and Leopold, 1978, Figure 10–8) through a reclassification of the cell values. Once a runoff curve was assigned to each cell, a rainfall surface was applied, and runoff generated for each cell. The total runoff into each cell was then calculated using the *flow accumulation* Grid function and the weighted runoff grid.

The AML code for surface erosion is based on the universal soil loss equation (USLE) as outlined by Dunne and Leopold (1978) where $A = RKLSCP$. A is soil loss in tonnes per hectare, R is the rainfall erosivity index, K is the soil erodibility index, L is the hillslope length factor, S is the hillslope gradient factor, C is the cropping management factor and P is the erosion control practice factor (Dunne and Leopold, 1978). This cell-based erosion potential was not routed or accumulated through the flow network, because the USLE does not account for sediment routing, and can only indicate the relative erosion potential for a cell. To check the model results, a local mass balance, i.e. the sum of all inputs minus the erosion potential, was calculated for each cell. This procedure produced the expected pattern of erosion on convex slopes and deposition on concave slopes.

Map algebra techniques were used to assemble all of the parameters except the P-factor for erosion control practice, which was assigned an intermediate value (Dunne and Leopold, 1978, Table 15–5, p. 530) of 0·5 for all cells. Assignment of values for the other factors is described in the following paragraphs and was based on soil, land cover/vegetation and rainfall data.

The most difficult parameter to derive for the USLE from a DEM (digital elevation model) is L. We used the *flow length* function in Grid to calculate the shortest distance from the ridge to each cell. According to Wischmeier and Schmidt (1978) [as described by Wilson and Gallant (1996)] the slope length can be normalized to the original 22·13 m slope for the USLE using the equation: $L\text{-factor} = (\text{flow length}/22\cdot13)^t$, where $t = 0\cdot5$ if the slope $b < 3°$; $t = 0\cdot4$ if the slope $1\cdot7 < b < 3$; $t = 0\cdot3$ if the slope $0\cdot6 < b < 1\cdot7$; and $t = 0\cdot2$ if the slope $b < 0\cdot6$.

In order to assign hydrological properties and K-values, soil properties had to be examined. The soil properties for Santa Cruz Island were based on research and digital data sets described by Jones and Grice (1993) and Butterworth *et al.* (1993). In particular, the digital soil map generated by Jones and Grice (1993, Figure 6 and Table 2) from data for geological substrate and vegetation was recreated using their digitized geological layer and the new land cover map generated from cluster analysis of the 20 October 1993 Landsat image, as described above. Soil types followed the assignments shown in Table 2 of Jones and Grice (1993) and were then divided by soil order into mollisols, entisols and inceptisols.

For the hydrological characterization, the thicker mollisols and river wash deposits were assigned high infiltration capacities relative to the entisols and inceptisols. Erodibility (K values) of the soils was based on considering that mollisols would tend to have more surface cohesion owing to more clay-sized particles and that the more poorly developed soil types would tend to have a higher proportion of coarser grained material that is more erodible from surface wash erosion. K values for the USLE (e.g. see Dunne and Leopold, 1978, Figure 15–17) of 0·3, 0·4 and 0·45 were assigned to the mollisols, inceptisols and entisols, respectively.

Without any detailed soil information for Santa Rosa Island, the geological units of Santa Cruz and Santa Rosa were compared. For those units that matched exactly, the same general soil assignment was made based on the vegetation and geological substrate. Tertiary intrusive rocks of Santa Rosa Island are described as bare surfaces in the highland region (Clark *et al.*, 1990) and were assigned a very low infiltration capacity and moderately high erodibility (0·45).

The rainfall surface that was used for both the runoff and erosion simulations for Santa Rosa and Santa Cruz islands was generated based on rainfall data from Santa Cruz Island. The analysis was designed to provide a generalized view of the types of conditions typical for a winter storm. Therefore, the four-day storm event leading up to the 9 February 1994 image was selected as a typical storm. This storm sequence generated 13 cm of rain at the low elevation rain gauge on Santa Cruz Island. According to observations by L. Laughrin (unpublished data), rainfall generally increases by 30% from the low rain gauge to the highest elevations of Santa Cruz island. As these are the best data currently available for both islands, a rainfall surface was linearly interpolated for both, starting with 13 cm (to approximate storm conditions for the time leading up to 9 February 1994) at the lowest island elevations and gradually increasing to 16·5 cm at approximately 800 m.

RESULTS

Four images representing different winter conditions are shown in Figure 3. The range of concentrations has not been calibrated using reference end-members as described by Mertes *et al.* (1993) because the optical characteristics of the coastal waters and the sediment in the plumes have not yet been fully characterized. However, a preliminary calibration has been developed, based on field measurements of total suspended matter measured the day before the image in Figure 3D was recorded. The field data yielded maximum total suspended matter concentrations near the mouth of the Santa Clara River of 70 mg/l, with rapid decreases in concentration as the distance from the shore increased (L. Mertes, unpublished data). Based on this range, the red colour in each image is estimated to have a concentration of the order of 100 mg/l, with a gradual decrease through the colours of the rainbow to <10 mg/l for the blue colours.

The spatial patterns of the plumes in these images represent a range of winter storm conditions. Figure 3A and 3B were recorded six weeks apart and show a distinctively different distribution of the high concentration features (yellows and reds). In the month prior to the recording of Figure 3A, only 9 cm of rain had fallen, with 8 cm of that total falling 11 days before the image was recorded. In the weeks between the two images, over 28 cm of rain fell, with 1·2 cm of rain falling 6 days before the image shown in Figure 3B was recorded. Figure 3A shows the Santa Clara–Ventura river plume as a red band hugging the area immediately offshore from the mouth. In contrast, the large plumes, of the order of 25 km wide, at the mouths of the Santa Ynez (see Figure 1 for river locations), in the north-west corner of Figure 3B, and the Santa Clara–Ventura River, in the south-east corner of Figure 3B, were generated from floods as the result of 28 cm of precipitation.

The storm total for the month prior to 9 February 1994 (Figure 3C) was approximately the same as for 4 January 1983. However, unlike for the 4 January 1983 conditions, there was nearly 4 cm of rain the day before the 1994 image was recorded. The narrow, high concentration, semi-circle (yellow and red in Figure 3C) may represent the jet flow of the plume as it entered the ocean before the ocean currents had a chance to diffuse and spread the plumes as in Figure 3A and 3B. Figure 3D was also recorded the day after a storm occurred, with approximately 3 cm of rain falling from 26–28 January. Although the spatial resolution of the AVHRR data at 1 km does not show the same detail as the Landsat data (80 m MSS and 30 m TM), the general pattern of the Santa Clara–Ventura plume location matches Figure 3B and 3C and its spatial location has been corroborated with measurements of suspended matter (L. Mertes, unpublished data).

The dominant spatial pattern apparent in Figure 3 is the variation in the patterns of the plumes at the mouths of the major rivers in the study area. A second-order pattern is that in all of the images the plumes surrounding Santa Rosa Island (SRI) are extensive and have a high concentration of particulates (yellows and reds). In contrast, there is little expression of plumes around Santa Cruz Island (SCI), which is of a comparable size to SRI. To determine the potential reasons for the differences in the island plume patterns several features were compared for each of the islands, including (i) morphometry, (ii) accumulated runoff, (iii) cell-based erosion potential and (iv) coastal geography.

The results of the morphometric analysis are shown in Figure 4 where a comparison of the drainage basin structure for each island for each order is presented. The statistics presented for each of the islands suggest that the islands are morphometrically similar in their first-, second-, and third-order watersheds. At the fourth order, the difference in the average stream length is substantial. Although the difference is not statistically significant owing to the high variance and low number of samples in the data (standard deviation of length for SRI is 2365 m and for SCI is 1636 m) the total stream length at the fourth order on SRI (58 200 m) is nearly twice as much as for SCI (31 220 m).

The next stage of analysis was to determine whether the accumulated runoff and cell-based erosion potential were substantially different. The results of the runoff calculation for the February 1994 storm are

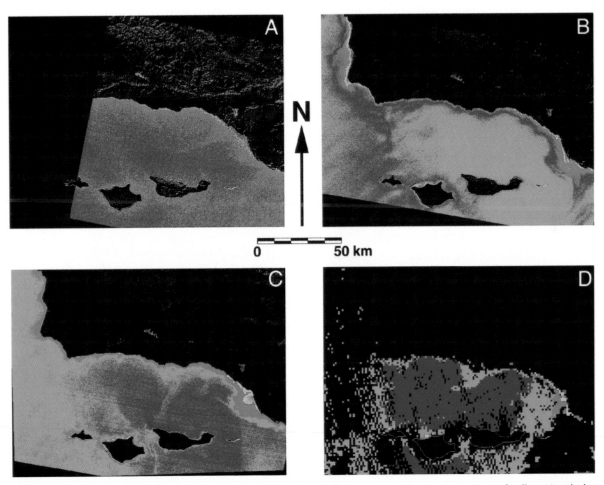

Figure 3. Sediment concentration patterns from remote sensing data. Four images show the changing patterns of sediment/particulate plumes in the Santa Barbara Channel. Colours indicate different particulate concentrations, with blues for concentrations less than 10 mg/l ranging through green to yellow to reds, with the highest concentrations of the order of 70–100 mg/l. All precipitation numbers from the Carpinteria gauge (Santa Barbara County Flood Control, unpublished data) (A) Landsat 3 MultiSpectral Scanner (MSS) image recorded 4 January 1983. 9 cm of rain fell in the preceding month. (B) Landsat 4 MSS image recorded 19 February 1983. 28 cm of rain fell in the preceding month. (C) Landsat 5 Thematic Mapper (TM) image recorded 9 February 1994. 10·5 cm of rain fell in the preceding month. (D) NOAA–AVHRR image recorded 29 January 1997. 4 cm of rain fell in the preceding month

(A)

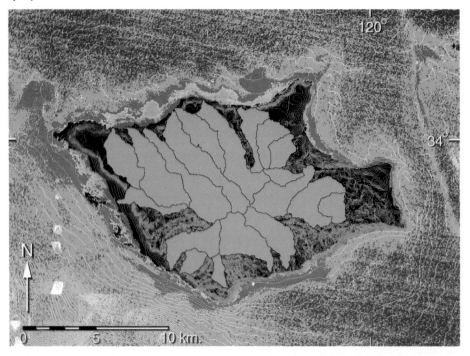

(B)

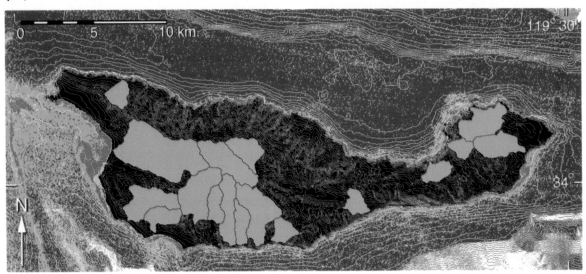

Figure 6. Watershed transport systems in relation to coastal geography for (A) Santa Rosa Island and (B) Santa Cruz Island. Figures include particulate concentrations in rainbow colours (see Figure 3C for colour description), bathymetric contours in white with 10 m contour intervals, watershed polygons for fourth-order drainage basins in gold and 30 m contours on land also in gold. The data sets were combined to emphasize the spatial correlations between the terrestrial watershed transport system, coastal plumes and the depth of the coastal waters

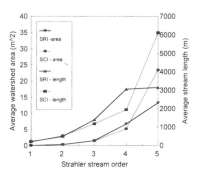

Figure 4. Morphometric characterization of Santa Rosa and Santa Cruz Islands. The numbers plotted for each order are: average watershed area and average stream length

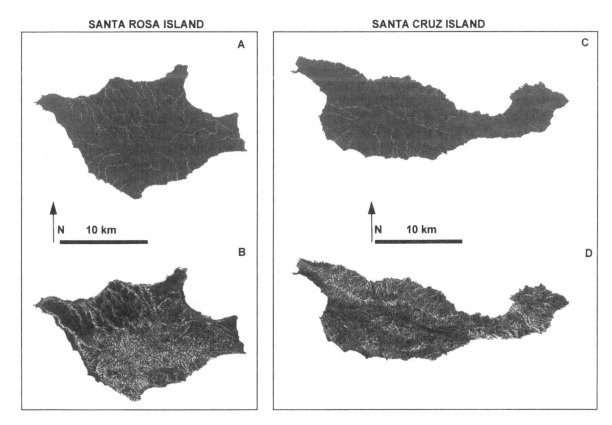

Figure 5. Water and sediment production for the four-day storm event prior to 9 February 1994. (A) and (C) Total volume of runoff accumulated for each cell for the storm event. The colour gradient is from white for high to grey for low total volume. (B) and (D) Potential erosion for individual cells (not accumulated or routed). Colour gradient is from white for high to dark grey for low erosion potential

shown in Figure 5A and 5C. The maximum accumulated runoff for a single channel at its mouth for SCI is $2\,220\,600$ m^3 (average of 0·3 m^3/s for the four-day storm event) and for SRI is $1\,188\,800$ m^3 (average of 0·15 m^3/s for the four-day storm event). The average rainfall–runoff percentage for SCI is 0·52, with a standard deviation of 0·13, and for SRI is 0·48, with a standard deviation of 0·10. The lower runoff volumes for SRI primarily reflect the fact that a linear interpolation scheme based on elevation was used for the rainfall surface. The maximum elevation on SRI is 475 m, versus 753 m for SCI (Power, 1980).

The pattern of erosion potential for each cell also showed comparable results for the two islands (Figure 5B and 5D). As expected, the largest erosion rates (shown as nearly white) are along the high, steep ridges of both islands. Mean values of 8 t/ha (SCI) and 6 t/ha (SRI) compare well with values ranging from 1·2 to 11 t/ha listed by Dunne and Leopold (1978, Table 15–1, p. 522) for southern California woodland and rangeland (originally reported by Krammes, 1960) and are well below the rate of 96 t/ha calculated from data reported by Keller *et al.* (1997).

DISCUSSION

The transfer of sediment to the ocean as exemplified by the development of plumes in the ocean after large storms is a function of the hydrology related to large storms, the efficiency of the watershed transfer of water and sediment and the spatial juxtaposition of watershed outlets and coastal geography. The remote sensing analysis described here and shown in Figure 3 represents synoptic views of the spatially and temporally dynamic processes controlling the transfer of particulates into the ocean by rivers. The first-order effects determined from analysis of the remote sensing data are that the largest watersheds, Santa Clara, Ventura and Santa Ynez, produce the largest sediment plumes. In addition, the El Niño-related storms in January and February 1983 contributed to the development of the largest observed plumes (Figure 3B).

The straightforward relationship between watershed size and plume size for the largest watersheds is not as clear for the smaller watersheds. For example, the plume patterns around the Northern Channel Islands appear to have been influenced by additional factors. It is informative to start with a morphometric analysis in order to determine possible reasons for the apparent difference between the plumes generated around Santa Rosa Island and Santa Cruz Island. The topography of Santa Rosa Island is dominated by a central highland formed along the Santa Rosa Island fault (Dibblee, 1982) from which all of the major drainage systems radiate and drain into the ocean (Figure 6A). The result of these island-scale structures is that the fourth-order drainage basins of Santa Rosa Island have relatively long stream lengths and the basins extend from the highest peaks to the ocean boundary (Figures 4 and 6). In contrast, Santa Cruz Island is cut in half by the Santa Cruz Island fault, which has produced a large interior valley flanked by ridges to the north and south (Dibblee, 1982). Therefore, many of the fourth-order drainage basins of Santa Cruz Island are confined to gentler parts of the landscape as the channels cross the low gradient central valley (Figure 6B).

With these structural differences in mind, the next step is to compare the potential for erosion for each island. The rates of erosion simulated for each island are similar. This pattern suggests that approximately equal amounts of sediment are being transferred to the channel system from the hillslopes of each island. Combining the structural patterns with the erosion rates yields the interpretation that the transfer of sediment from the ridges and hillslopes of Santa Cruz Island to the ocean is less efficient because of long-term storage in the central valley than the more direct transfer achieved by the radial drainage of Santa Rosa Island.

The extensive plumes around Santa Rosa Island appear to be a function not only of the apparently more efficient watershed transfer system, but also of the relative position of watershed outlets and shallow coastal areas. For both of the islands, Figure 6 shows the fourth-order watershed boundaries in gold, the plume patterns for the February 1994 event in rainbow colours, the 10 m bathymetric contours in white and the 30 m contours on land in gold. In general, the most extensive plumes near Santa Rosa Island are in areas that are less than 20 m deep, especially on the north and west sides of the island. The only large plumes near Santa Cruz Island are in the shallow bay on the west end. Hence, a further influence on the plumes is that material transported and perhaps deposited in the shallow coastal waters may be susceptible to resuspension, even after the terrestrial inputs have ceased.

The susceptibility to resuspension is a function of wave energy, particle size and water depth. These factors have been combined into Dean's parameter, which is a dimensionless index of sediment fall velocity and wave height and period (Short and Wright, 1983). A morphodynamic classification has been developed for beaches based on Dean's parameter and essentially it expresses the ability of waves to move and resuspend sediment at that beach. According to J. Dugan (personal communication) many of the beaches of the Northern Channel Islands, Santa Barbara, Ventura and Los Angeles counties are primarily intermediate morphodynamic beach types. Intermediate beaches are prone to resuspension and significant exchange of sediment between the beach face, the surf zone and the near-shore zone. The shallow, exposed beaches and shallow coastal regions of the north and west shores of Santa Rosa Island may be particularly susceptible to resuspension of sediment. In contrast, the bathymetry around SCI is, on average, 10–20 m deeper and there is very little evidence for resuspension of sediments around the island (Figure 6).

CONCLUSIONS

In this study we analysed synoptic views of the land–marine interface of the Santa Barbara Basin with remote sensing data recorded for a range of winter storm conditions. Based on these data it was determined that the first-order controls on the spatial patterns and extent of the coastal plumes are (i) the size of the coastal watershed and (ii) storm size. Plumes associated with the smaller drainage basins did not show such a simple relationship, especially for the Northern Channel Islands that bound the southern edge of the basin.

In order to investigate the influences on the smaller plumes, several GIS-based analyses were completed for two of the islands, Santa Rosa and Santa Cruz. These analyses included (i) watershed morphometry, (ii) accumulated runoff, (iii) cell-based erosion potential and (iv) near-shore bathymetry. Based on these data and simulations it was concluded that Santa Rosa Island has an efficient sediment transport system from the ridges to the ocean because of a radial drainage pattern. In addition, many of the largest watersheds on Santa Rosa Island have outlets in shallow coastal waters. Sediment may be subject to resuspension in these shallow areas, thus creating the appearance of larger sediment plumes than might be expected for the small watersheds. In contrast, most of the large drainage basins on Santa Cruz Island cross the low gradient central valley and enter deep coastal waters.

ACKNOWLEDGEMENTS

This research was supported primarily by funding from the NOAA Channel Islands National Marine Sanctuary, the UCSB Instructional Development Program, NOAA and the Office of Naval Research. Additional funding was provided by the Channel Islands National Park, University of California Natural Reserve System, a UCSB Undergraduate Research–Genesis Award (to M. H.) and the State of California Office of Oil Spill Prevention. Discussions with D. Maidment, J. Wilson, B. Lees and M. Cobb contributed enormously throughout the development of this paper. Reviews by D. Montgomery and an anonymous reviewer improved the final manuscript. E. Fields provided the calibrated AVHRR data set. K. Schwemm provided the digital vegetation map of Santa Rosa Island. This paper is dedicated to the memory of Waifun (Olivia) Au Yeung who always asked the best questions.

REFERENCES

Behl, R. J. and Kennett, J. P. 1996. 'Brief interstadial events in the Santa Barbara Basin, NE Pacific, during the past 60 yrs', *Nature*, **379**, 243–246.

Best, D. W. and Keller, E. A. 1986. 'Sediment storage and routing in a steep boulder-bed rock-controlled channel', in DeVries, J. J. (ed.), *Proceedings of Chaparral Ecosystems Research Conference*. California Water Resources Center, Sacramento, California.

Brumbaugh, R. W. 1980. 'Recent geomorphic and vegetal dynamics on Santa Cruz Island, California', in Power, D. M. (ed.), *The California Islands — Proceedings of a Multidisciplinary Symposium*. Santa Barbara Museum of Natural History, Santa Barbara. pp. 139–158.

Butterworth, J. P., Jones, J. A., and Jones, S. 1993. 'Soil forming factors, morphology, and classification — Santa Cruz Island, California', in Hochberg, F. G. (ed.), *Third California Islands Symposium — Recent Advances in Research on the California Islands*. Santa Barbara Museum of Natural History, Santa Barbara. pp. 39–44.

Cole, K. L. and Liu, G-W., 1994. 'Holocene paleoecology of an estuary on Santa Rosa Island, California', *Quat. Res.*, **41**, 326–335.

Clark, R. A., Halvorson, W. L., Sawdo, A. A., and Danielsen, K. C. 1990. 'Plant communities of Santa Rosa Island, Channel Islands National Park', *Cooperative National Park, Resources Studies Unit Technical Report 42*. University of California, Davis, p. 93.

Dibblee, T., Jr 1982. 'Geology of the Channel Islands, Southern California', *Geology and Mineral Wealth of the California Transverse Ranges.* South Coast Geological Society. pp. 27–39, Santa Barbara, California.

Drake, J. B., Kolpack, R. L., and Fisher, P. J. 1972. 'Sediment transport in the Santa Barbara Oxnard Shelf, Santa Barbara Channel, California', in Swift, D. P., Duane, D., and Pilkey, O. H. (eds), *Shelf Sediment Transport*: Dowden, Hutchinson, & Ross, Stroudsburg, Pennsylvania. pp. 307–329.

Dunne, T. and Leopold, L. 1978. *Water in Environmental Planning.* W. H. Freeman and Co., New York. 818 pp.

Fan, P-F. 1976. 'Recent silts in the Santa Clara River drainage basin, Southern California: a mineralogical investigation of their origin and evolution', *J. Sediment. Petrol.*, **46**, 802–812.

Ferren, W. R., Jr, Capelli, M. H., Parikh, A., Magney, K. D., Clark, K., and Haller, J. R. 1990. 'Botanical resources at Emma Wood State Beach and the Ventura River Estuary, California: inventory and management', *The Herbarium–Department of Biological Sciences Environmental Report 15.* University of California, Santa Barbara.

Florsheim, J. L., Keller, E. A., and Best, D. W. 1991. 'Fluvial sediment transport in response to moderate storm flows following chaparral wildfire, Ventura County, southern California', *Geol. Soc. Am. Bull.*, **103**, 504–511.

Haston, L. and Michaelsen, J. 1994. 'Long-term central coastal California precipitation variability and relationships to El-Nino-Southern oscillation', *J. Clim.*, **7**, 1373–1387.

Holligan, P. M. and Reiners, W. A. 1992. 'Predicting the responses of the coastal zone to global change', *Adv. Ecol. Res.*, **22**, 211–255.

Ingersoll, R. V., Kretchmer, A. G., and Valles, P. K. 1993. 'The effect of sampling scale on actualistic sandstone petrofacies', *Sedimentology*, **40**, 937–953.

Jones, J. and Grice, D. 1993. 'A computer generated soils map of Santa Cruz Island, California', in Hochberg, F. G. (ed.), *Third California Islands Symposium — Recent Advances in Research on the California Islands.* Santa Barbara Museum of Natural History, Santa Barbara. pp. 45–56.

Keller, E. A. and Capelli, M. H. 1992. 'Ventura River flood of February 1992 — a lesson ignored', *Wat. Resour. Bull.*, **28**, 813–832.

Keller, E. A., Johnson, D. L., Rockwell, T. K., Clark, M. N., and Dembroff, G. R. 1982. 'Tectonic geomorphology of the Ventura, Ojai, and Santa Paula areas, Western Transverse Ranges in Southern California', Geological Society of America Guidebook, Denver, Colorado, 33 pp.

Keller, E. A., Valentine, D. W., and Gibbs, D. R. 1997. 'Hydrological response of small watersheds following the Southern California Painted Cave Fire of June, 1990', *Hydrol. Process.*, **11**, 401–414.

Kennedy, J. A. and Brassell, S. C. 1992. 'Molecular stratigraphy of the Santa-Barbara Basin — comparison with historical records of annual climate change', *Org. Geochem.*, **19**, 235–244.

Kolpack, R. L. and Drake, D. E. 1984. 'Transport of clays in the eastern part of Santa Barbara Channel, California', *Geo-Marine Lett.*, **4**, 191–916.

Kondolf, G. M. and Keller, E. A. 1991. 'California watersheds at the urban interface', *Proceedings of the 3rd Biennial Watershed Conference, Ontario, Canada.* pp. 27–40.

Krammes, J. S. 1960. 'Erosion from mountain sideslopes after fire in southern California', *US Forest Service Pacific Southwest Forest and Range Experiment Station Research Note 171.* USFS, Berkeley.

Maidment, D. R. 1995. 'GIS and hydrology', *User Notes for 15th Annual ESRI User Conference, Palm Springs, CA.*

Marsaglia, K. M., Rimkus, K. C., and Behl, R. J. 1995. 'Provenance of sand deposited in the Santa Barbara Basin at site 893 during the last 155,000 years', in Kennett, J. P., Bauldauf, J. G., and Lyle, M. (eds), *Proceedings of the Ocean Drilling Program — Scientific Results*, Vol. 146 (part II). College Station, Texas. pp. 61–75.

Mertes, L. A. K., Smith, M. O., and Adams, J. B. 1993. 'Estimating sediment concentrations in surface waters of the Amazon River wetlands from Landsat images', *Rem. Sens. Environ.*, **43**, 281–301.

Milliman, J. D. and Syvitski, J. P. M. 1992. 'Geomorphic/tectonic control of sediment discharge to the ocean: the importance of small mountainous rivers', *J. Geol.*, **100**, 525–544.

Minnich, R. A. 1980. 'Vegetation of Santa Cruz and Santa Catalina Islands', in Power, D. M. (ed.), *The California Islands — Proceedings of a Multidisciplinary Symposium.* Santa Barbara Museum of Natural History, Santa Barbara. pp. 123–138.

Mueller, J. L. and Austin, R. W. 1995. 'Ocean optics protocols for SeaWiFS validation, Revision 1', *SeaWifs Technical Report Series.* NASA Technical Memorandum 104566, vol. 25. Washington, D.C..

NOAA (National Oceanic and Atmospheric Administration) 1996. *Geophysical Data System for Hydrographic Survey Data.* National Geophysical Data Center, Boulder.

Patterson, R. H. 1979. 'Tectonic geomorphology and neotectonics of the Santa Cruz Island Fault, Santa Barbara County, California', *PhD Thesis*, Santa Barbara, University of California. 141 pp.

Power, D. M. (ed.) 1980. 'The California Islands', *Proceedings of a Multidisciplinary Symposium.* Santa Barbara Museum of Natural History, Santa Barbara. p. 3.

Putnam, W. C. 1942. 'Geomorphology of the Ventura Region, California', *Geol. Soc. Am. Bull.*, **53**, 691–754.

Renwick, W. 1982. 'Landslide morphology and processes on Santa Cruz Island, California', *Geograf. Annal.*, **64A**, 149–159.

Rice, R. M. 1982. 'Sedimentation in the chaparral: how do you handle unusual events?', in Swanson, F. J., Janda, R. J., Dunne, T., and Swanston, D. N. (Eds), *Sediment Budgets and Routing in Forested Drainage Basins*, US Forest Service, General Technical Report PNW-141. United States Forest Service, Pacific Northwest Research Station, Corvallis, Oregon. pp. 39–49.

Rice, R. M., Gorsline, D. S., and Osborne, R. H. 1976. 'Relationships between sand input from rivers and the composition of sands from the beaches of Southern California', *Sedimentology*, **23**, 689–703.

Rockwell, T. K., Keller, E. A., Clark, M. N., and Johnson, D. L. 1984. 'Chronology and rates of faulting of Ventura River terraces, California', *Geol. Soc. Am. Bull.*, **95**, 1466–1474.

Schimmelman, A., Lange, C. B., and Berger, W. H. 1990. 'Climatically-controlled marker layers in Santa-Barbara Basin sediments and fine-scale core-to-core correlation', *Limnol. Oceanogr.*, **35**, 165–173.

Scott, K. M. and Williams, R. P. 1978. 'Erosion and sediment yields in the Transverse Ranges, Southern California, U.S.', *United States Geological Survey Professional Paper 1030*. 37 pp.

Short, A. D. and Wright, L. D. 1983. 'Physical variability of sandy beaches', in McLachlan A. and Erasmus, T. (Eds), *Sandy Beaches as Ecosystems*. Dr. W. Junk, The Hague. pp. 133–144.

Taylor, B. D. 1983. 'Sediment yields in coastal southern California', *ASCE J. Hydraul. Engng*, **109**, 71–85.

Weaver, D. W., Doerner, D. P., and Nolf, B. 1969. 'Geology of the northern Channel Islands (California)', *AAPG Pacific Section, Special Publication*. American Association of Petroleum Geologists. 200 pp.

Wilson, J. and Gallant, J. 1996. 'EROS: a grid-based program for estimating spatially-distributed erosion indices', *Comp. Geosci.*, **22**, 707–712.

12

MORPHOLOGICAL AND ECOLOGICAL CHANGE ON A MEANDER BEND: THE ROLE OF HYDROLOGICAL PROCESSES AND THE APPLICATION OF GIS

A. M. GURNELL,* M. BICKERTON, P. ANGOLD, D. BELL, I. MORRISSEY, G. E. PETTS AND J. SADLER

School of Geography, University of Birmingham, Birmingham B15 2TT, UK

ABSTRACT

Environmental change induced by hydrological processes can often be quite small. This paper illustrates subtle changes in river planform on a single meander bend of the lower River Dee, Wales, and the significance of those changes for the ecology of the present riparian zone. Geographical information systems (GIS) are shown to provide an excellent framework for integrating historical and contemporary information from different sources, and for quantifying possible transcription errors so that true environmental associations and relatively small changes, in the context of the spatial scales of the sources, can be identified with confidence. As a result, functional units within the riparian zone of a regulated river can be defined and mapped.

KEY WORDS   historical data sources; air photographs; geographical information systems; channel planform change; hydroecology

INTRODUCTION

River margins are ecotonal environments characterized by high ecological diversity and productivity (Petts, 1990). They are strongly influenced by hydrological processes, especially the frequency, duration and timing of inundation. The range and distribution of ecological patches also reflects the geomorphological dynamics. These patches comprise functional units, each being characterized by a typical plant and animal community that is indicative of habitat conditions at the site (Petts and Amoros, 1996). These functional units are generally arranged along topographic gradients (or toposequences). They also form chronosequences: successional sequences influenced by sedimentation, organic matter accumulation, population development and species replacement. These functional units may be grouped into functional sets associated with different geomorphological settings. One example of this is the transformation around a meander from 'point' to 'counterpoint' depositional environments. The point bar/berm is characterized by a different set of functional units than the silty counterpoint bar/berm. In contrast to the classic point bar/berm, with sandy alluvium overlying gravel, counterpoint bar/berms are slackwater deposits characterized by organic silts (Shi *et al.*, submitted). One approach to integrating accurately historical and contemporary information in these types of environment is to investigate the structure and dynamics of the river margin within a GIS (geographical information system).

---

* Correspondence to: A. M. Gurnell, School of Geography, University of Birmingham, Birmingham B15 2TT, UK.

Contract grant sponsor: European Union.
Contract grant number: ENV4-CT95-0062.

This paper examines the development of a point to counterpoint berm on the River Dee, Wales, and then explores the interactions between morphology and ecology on the counterpoint section of the berm. It provides an example of the use of GIS in identifying the role of hydrological processes within the riparian zone. Whilst the research uses only simple GIS functions, particularly the overlaying of information derived from different sources and representing different properties of the riparian zone, it makes use of information on the errors involved in data transcription to define patterns and changes that are most likely to represent genuine environmental changes and correlations. Specifically, GIS is used as a framework for identifying changes over two time-scales, decadal and seasonal.

First, approximately decadal changes are characterized over the period 1897–1992, a time of increasing regulation of the river flows and thus a gradually changing hydrological regime. Analysis of the planform evolution of a berm and of its vegetation cover over a period of *c.* 100 years is accomplished within a GIS using information from map and aerial photograph sources.

Secondly, seasonal changes are investigated by considering the interaction between the berm as represented by a digital elevation model (DEM), and the water level frequency distribution at the annual time-scale. At this time-scale, there is negligible change in the berm size and morphology, but, as a result of its interaction with the discharge regime of the river, the morphology controls the distribution of areas of different inundation frequency. Areas characterized by different inundation environments support the development of different plant and animal communities.

THE STUDY AREA

The catchment area of the River Dee to Chester Weir (Figure 1) is 1817 km². It generates a mean river flow of 37 m³ s⁻¹ (Lambert, 1988). Since the early nineteenth century the river has been subject to increasing flow regulation. The main changes in flow regulation relate to the closure of the impounding Alwen Reservoir (capacity 15 million m³) during the 1920s; the regulation of Bala Lake in the 1950s (which resulted in some 18 million m³ of controllable water storage); the completion of the regulating reservoir Llyn Celyn in 1965 (capacity 81 million m³); and the completion in 1979 of the regulating reservoir Llyn Brenig (capacity 60 million m³). Thus, reservoir storage within the Dee catchment has developed strongly within this century, and has been particularly marked over the last 50 years. Regulation has had a significant influence on the river flow regime. Figure 2 illustrates changes in some annual summary statistics of the monthly instantaneous maximum flows, 1938–1992, at the Erbistock/Manley Hall gauging stations. These changes include a significant decline throughout the period in the annual maximum, upper quartile, mean and standard deviation (Figure 2A and B), and a steady increase since the 1960s in the annual minimum of the monthly instantaneous maximum flows (Figure 2C). The mean annual flood for the period 1965–1992 (i.e. after the closure of the largest volume of controllable water storage; the Llyn Celyn reservoir) is 231 m³ s⁻¹, which is 14% smaller than the 269 m³ s⁻¹ mean annual flood for the period 1946–1963 (i.e. the period after the regulation of Bala Lake).

The study site is located a short distance downstream from the Erbistock/Manley Hall gauging stations (Figure 1) and upstream of the main water abstractions. Previous research (Gurnell *et al.*, 1994; Gurnell, 1997a,b) has illustrated complex changes throughout an 18 km reach of the River Dee (Figure 1c), commencing upstream at the present study site and extending downstream across the fluvial–tidal transition. The timing of these changes and their translation downstream has been shown to correlate with the progressive regulation of the river (Gurnell *et al.*, 1994; Gurnell, 1997a). The channel has narrowed, and, in the upstream portion including the present study site, the narrowing since the late 1940s has been achieved by the deposition of low-level sedimentary berms (Shi *et al.*, submitted). Analysis of actual and naturalized (i.e. estimates of flows in the absence of regulation and abstraction) flow frequency data for the period 1970–1989, coupled with grain size-specific, bedload sediment rating curves (derived using Bagnold's 1966, 1980 stream power function), suggests enhanced sediment transport under the actual (regulated) flow regime

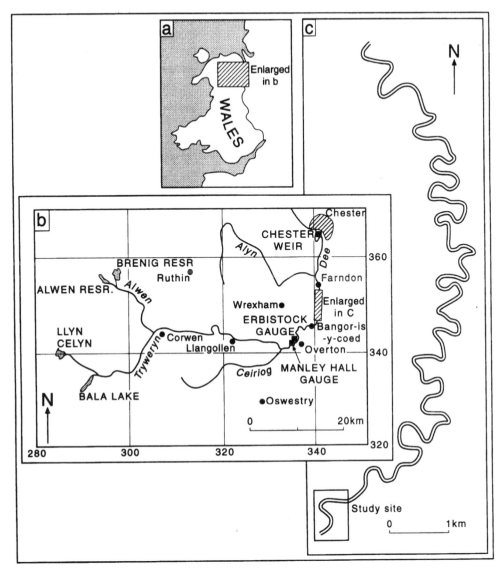

Figure 1. Location of the study site

for flows in the range 5–100 m³ s⁻¹, with the midpoint falling in the range 47·5–70·0 m³ s⁻¹ for all of the years analysed, but a reduction in sediment transport outside this flow range (Gurnell, 1997b).

CHANGES IN BERM EXTENT, MORPHOLOGY AND VEGETATION COVER 1897–1992

Estimation of river channel planform change within a GIS

Historical map and air photograph sources are useful for investigating river planform change if the amount of planform change exceeds the spatial errors inherent in the sources and in the methods of transcription used to support the analysis of the map-derived information (Downward, 1995).

The accuracy with which planform change can be identified from map sources depends upon the map scale, the conventions used for identifying the position of river banks and the accuracy with which that position is surveyed and plotted on the final map sheet. Furthermore, the level of temporal detail of planform

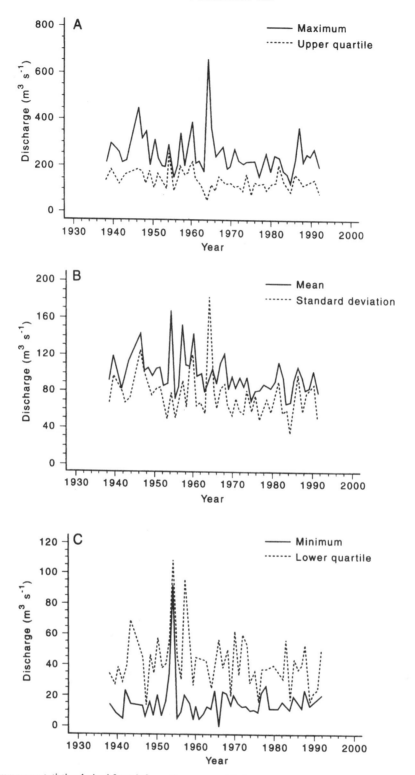

Figure 2. Annual summary statistics derived from information on monthly instantaneous maximum flows at the Erbistock/Manley Hall gauging stations, 1938–1992

change identified from map sources depends upon the number of historical map surveys that are available for the study river, and the precision with which the date of the survey of the channel can be identified (Downward *et al.*, 1994; Gurnell *et al.*, 1994). In Britain, river planform change over the last 100–150 years can be explored using Ordnance Survey 1 : 10 560 and 1 : 10 000 scale maps. These maps give reasonably high spatial detail concerning river margin position. They represent the level of 'normal' winter flow following standard Ordnance Survey practice (Harley, 1975) to identify the position of river channel boundaries. In essence, this equates to the morphological upper limit of the banks of the active channel at the time of survey. Maps at this scale also have the advantage of fairly frequent repeat surveys to provide information on channel planform change.

Air photographs contain more information that is relevant to the interpretation of processes of channel change than can be obtained from topographic maps (Milton *et al.*, 1995). For example, information on bank vegetation cover, detailed in-channel and riparian morphological features and the pattern of sediment deposition can be identified (Gurnell, 1997a). Moreover, the frequency of repeat air photograph cover is usually higher than topographic map resurveys. However, information obtained from air photographs requires even more careful handling than that from topographic maps if errors in data extraction and registration are to be minimized. Particular problems are the planimetric distortion of the photographs and the interpretation of morphological and other boundaries of natural features that grade into one another across fuzzy boundaries on the photographs.

Here, river channel change is explored along a *c.* 1 km length of the River Dee, Wales (Figure 1), using both map and air photograph sources. Information extracted from three sets of topographic maps (1897, 1949, 1979) and six sets of air photographs (1946, 1951, 1966, 1974, 1985 and 1992) is integrated and analysed within a GIS. The GIS software used was SPANS GIS for OS2, version 5·3 (Intera Tydac Technologies Inc., 1993). The three map sources and the most recent air photographs provide information on planform change over the last century (1897–1992), whereas the six air photograph sources provide more detailed information for the last *c.* 50 years (1946–1992) (Table I).

A GIS-based approach for handling historical information on channel planform has a number of advantages over manual procedures (Gurnell *et al.*, 1994). Of particular importance in the present context is the ability to register all of the extracted information to a common base and to derive estimates of the degree to which differences in the character of the river at different dates are likely to be a result of genuine geomorphological change rather than errors accumulated in the handling of the data. Full details of the methods employed in handling the map (Gurnell *et al.*, 1994) and air photograph information (Gurnell,

Table I. Historical sources analysed

Date used in analysis	Edition date/Sortie	Survey dates	Comments/Scale
Ordnance Survey maps			
1897	1900	Denbeigh 1897–1898 Cheshire 1897	1 : 10 560 complete coverage
1949	1954	1949	1 : 10 560 revised after 1930, major changes revised 1949
1979	1975–1979	1972, revised 1976–1979	1 : 10 000
Air photographs			
1992	Geonex	27·06·92	1 : 10 000
1946	106G/UK 1454	02·04·46	1 : 9840
1951	540/491	11·05·51	1 : 10 000
1966	OS 66-81	29·05·66	1 : 23 000 (analysis based on ~1 : 11 500 scale prints)
1974	Meridian airmaps	14·06·74	1 : 10 000
1985	J. A. Storey	01·06·85	1 : 10 000
1992	Geonex	26·06·92	1 : 10 000

1997a) are provided elsewhere. In brief, standardised methods were developed for transcription and registration of information from each of the data sources to a common scale and map projection, utilizing the Ordnance Survey National Grid.

For the map sources, information on river bank position was digitized at a varying interval selected to reflect fully the bank planform. The resultant vectors were registered within the GIS using a large sample of control points and a linear least-squares transformation. After transformation the displacement of each control point with respect to the National Grid was estimated by taking the square root of the sum of the squares of the residual easting and northing values. Although the distributions of displacements were truncated at 0·0 m (only positive displacements are possible), they were otherwise approximately normally distributed. In the case of the most recent (1979) map sheets, the National Grid was overprinted and so grid intersections were used as control points. The mean and standard deviation of the displacements were 1·36 and 0·59 m, respectively, from a sample size of 39. The 1979 map sheet was used as the common base for defining coordinates and thus registering information from all other historical sources (maps and photographs). The largest displacements associated with map sources were identified for the earliest map sheet. Here, fixed features were identified and digitized from a second-generation photocopy of the map and were registered to the common base using their coordinates on the 1979 map sheet. A mean displacement of 4·48 m and a standard deviation of 1·34 m were found for a sample of 38 control points. In addition to the map sources, the 1992 air photographs were geometrically corrected and registered using ortho-imaging software, so that the 1992 normal winter water level (used to define bank position on Ordnance Survey maps) could be interpreted using morphological criteria and digitized on-screen for comparison with the map information. This approach was time-consuming and, given the relatively subdued vertical relief of the Dee floodplain, a simpler approach, which is detailed below, was developed subsequently for extracting comparative planform information from the six sets of air photographs.

A number of strategies were adopted to maximize the comparability of the information extracted from air photographs of the same section of channel at different dates. Air photograph cover of similar scale was employed (Table I) and was interpreted by the same operator. The limits of the channel were not based on morphological criteria but on the channel bank vegetation limit, which can be more readily identified from photographs. The vegetation limit was digitized directly from the air photographs and the resultant vectors were subsequently transformed and registered to the National Grid. A set of standard control points, identified on the 1979, 1:10 000 scale, Ordnance Survey map cover, was used to register information from each of the photographs to the National Grid. The number of control points used for registering individual photographs was typically in the range of 15–25 points. A standard non-linear least-squares transformation was employed to accommodate the horizontal distortions in the photographs when transforming and registering information. A sample of 82 residuals for the 1946 set of air photographs generated the largest residual displacements, with an average of 3·6 m and a standard deviation of 2·1 m.

The magnitude of the residuals resulting from transformation and registration of both the maps and air photographs quoted above (and in each case relating to the oldest and thus most problematic data sources) suggests that observed changes in channel position in excess of 5 m are very likely to represent genuine channel change. The following description of channel evolution will focus on positional changes that exceed 10 m.

Planform change estimated from map and air photograph sources

Figures 3 and 4 plot the changing position of the channel boundaries as represented by the normal winter water level (a morphologically defined boundary) and the limit of perennial vegetation, respectively. Figures 3 and 4 only plot boundaries from selected maps and air photographs to highlight the major changes that have occurred. Reference to information for other dates will be made where relevant to the discussion.

Figure 3 illustrates migration of the channel towards the east and channel narrowing over the last 100 years. Channel migration and narrowing between 1897 and 1949 resulted in a zone of deposition over 50 m wide at the maximum along the left bank of the river. This zone now forms part of the floodplain.

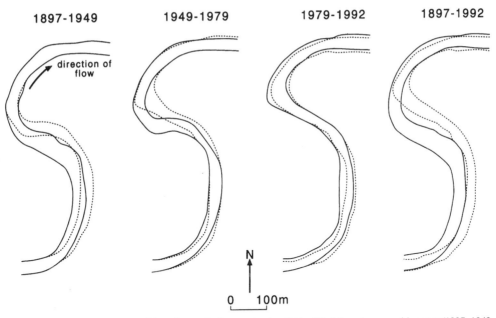

Figure 3. Changes in channel boundary positions (normal winter water level) identified from topographic maps (1897, 1949, 1979) and air photographs (1992). In each map the solid lines represent the earlier date and the dashed lines represent the later date

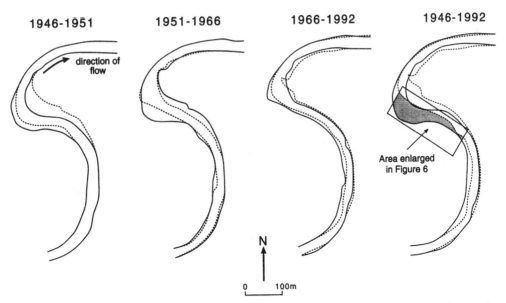

Figure 4. Changes in channel boundary positions (vegetation limit) identified from air photographs. In each map the solid lines represent the earlier date and the dashed lines represent the later date

Migration of the southern and central part of the channel (1949–1979), and general migration and channel narrowing (1979–1992) have resulted in the development of a low berm extending up to 50 m in width along the left bank of the river and *c.* 1·5–2 m below the level of the floodplain. Thus, whereas early channel migration was associated with extension of the floodplain, changes since the late 1940s (i.e. since the commencement of the recent phase of increasing regulation of river flows) has been associated with berm construction. Berm construction is characteristic of many regulated rivers (Shi *et al.*, submitted). Figure 4 provides greater detail of the development of the berm over the last 50 years. The counterpoint section of the berm (in the northern half of Figures 3 and 4) developed progressively from 1946, reaching its present limit in 1974. The point berm (in the southern half of Figures 3 and 4) started to build between 1951 and 1966 and then gradually extended to 1992. Thus, the general pattern of growth identified by overlaying channel boundary positions from the historical sources is as follows.

(i)   1897–1940s. Extension of the left bank floodplain through channel migration.

(ii) 1940s–1970s. Development of the counterpoint berm to its present extent; patchy initiation of the point berm.

(iii) 1970s–1990s. Development of point and counterpoint berms and their merger along the entire left bank and steady extension eastwards of the central and southern sections of the berm throughout the period.

Figure 5 illustrates the main zones of vegetation on the extended floodplain and berm that are identifiable from the air photographs. Zone I represents the 1897–1940s extension of the floodplain. In 1946 this area had an unstable poached/broken cover which had developed a stable pasture community by 1966 and was indistinguishable from the surrounding floodplain pasture by 1985. Zone I forms an extension of the floodplain and is built at approximately the same elevation. Zones II–V represent areas of the lower level berm, which has been constructed since the late 1940s. Zone II represents the counterpoint section of the berm, which was deposited between the 1940s and 1970s. In 1966, the first development of *Salix* marked the landward margin of the berm. Subsequent photographs illustrate an expansion of the woodland canopy, with a corresponding understorey development of tall herb vegetation to reach its present extent by 1985. Zone III includes the patchy initiation of the point berm during the 1940s–1970s. This zone included a narrow linear strip of herbaceous riparian vegetation in 1946, which developed into a linear fringe of *Salix* between 1966 and 1985. In the 1940s, a narrow linear herbaceous riparian strip marked the top of the bank in zone IV. A sparse line of *Salix* colonized the strip by 1966 and increased, spreading on to the developing berm in subsequent photographs so that by 1992 the canopy, including other species, covered approximately 75% of the berm surface. Evolution of the berm in zone V was associated with a narrow strip of bare ground

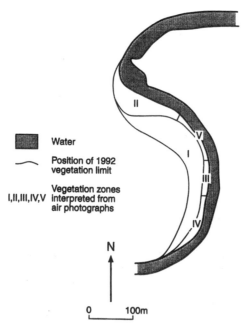

Figure 5. Vegetation zones associated with the development of the berm. For explanation of I to V, see text

in 1946, developing to a narrow, linear strip of tall herbaceous riparian vegetation by 1966. As the berm extended, a linear fringe of *Salix* developed after 1966.

SEASONAL BERM INUNDATION PATTERNS AND THEIR ECOLOGICAL CONSEQUENCES

This section considers interactions between the present morphology of the counterpoint section of the berm and the frequency of inundation of the berm surface. It also assesses the significance of these interactions for the ecology of the berm. Spatially continuous information on the three-dimensional form of the berm and the extent and frequency of different water levels within the river can be combined with information derived from point locations on the vegetation composition and the composition of the invertebrate fauna to assess the affect of physical processes on the ecology of the berm.

Berm morphology, flow duration and inundation frequency

Figure 6 illustrates the current morphology of the counterpoint section of the berm, including zone II and a part of zone V (Figure 5), as illustrated by a DEM. An analysis of the last 20 years of daily mean flows at the Erbistock/Manley Hall gauging stations in relation to local water levels on the berm allows the mapping of inundation percentiles across the berm surface based upon the 20-year flow duration curve (Figure 6). The mapped intersections between inundation percentiles and the berm morphology illustrate that inundation of the entire berm surface occurs within the range of the 25 to 10 flow percentiles. This is equivalent to a discharge range of 38–74 m^3 s^{-1}, and is very close to the 47·5–70 m^3 s^{-1} estimates based on previous simple modelling approaches (Gurnell, 1997b). Although zone V slopes gently towards the river, zone II contains topographic hollows which relate to the early period (1940s–1970s) of rapid lateral extension of the berm. Indeed, it is likely that part of the initial development of the berm involved the attachment of a marginal bar. Not only does the berm morphology support such an origin, but this process has been observed as a part of berm development on nearby sections of the Dee (Shi *et al.*, submitted). Furthermore, the 15-year gap between the 1951 and 1966 air photographs would have been sufficient time for such a process to occur. Although surface inundation of these hollows cannot occur until water spills across the higher surface near

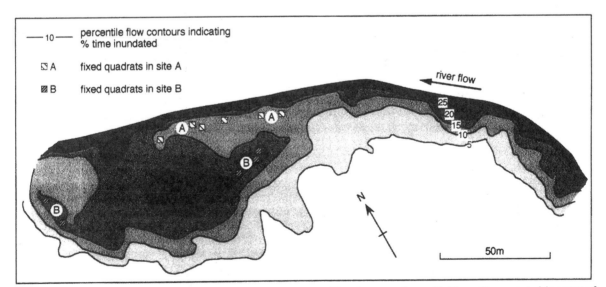

Figure 6. Morphology of the counterpoint section of the berm (zone II and a part of zone V, Figure 5) represented in terms of inundation frequency percentiles. Sites A and B are the areas within which ecological surveys (the results of which are presented in Figure 7) were undertaken. Vegetation surveys utilized random and fixed quadrats (the locations of the latter are indicated) and invertebrates were collected in pitfall traps located close to the fixed quadrats

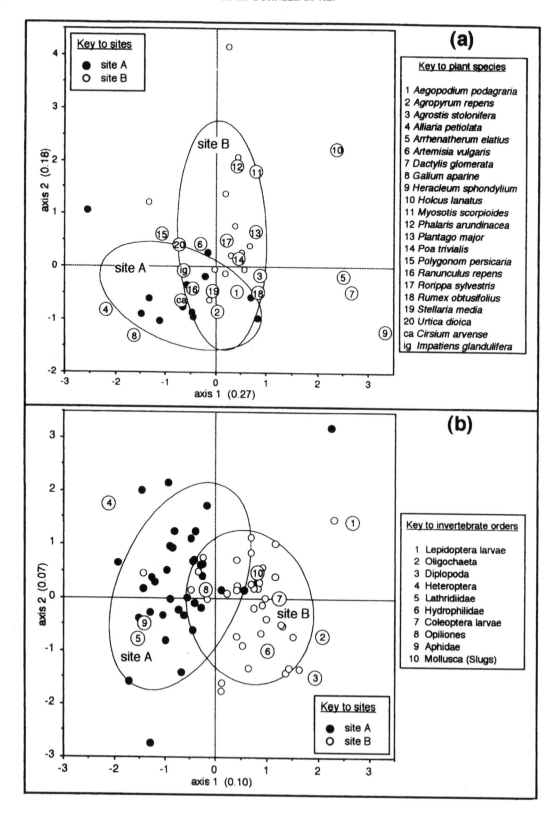

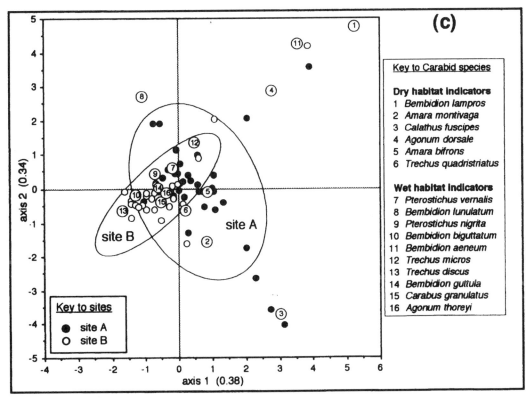

Figure 7. Ordination plots based on axis 1 and 2 scores from detrended correspondence analysis of: (a) plant species; (b) all invertebrate taxa; and (c) carabid beetle species. Sample points from site A (occasionally flooded) and site B (frequently flooded) are distinguished by black and white circles, respectively. Particular species/taxa are represented by numbered open circles

the channel, seepage of water through the berm allows water levels in the hollows to mirror closely those in the main channel.

Associations between inundation frequency and berm ecology: vegetation and invertebrate communities

A variety of ecological information has been collected for a network of locations within two sites across the berm surface: site A (occasionally flooded, bounded by the 10–15 inundation percentiles) and site B (frequently flooded, bounded by the 15–25 inundation percentiles) (Figure 6). Although these represent an apparently small difference in inundation frequency, the two sites can be shown to have clear differences in their ecological characteristics. For example, Figure 7 illustrates ordination plots for plant species (Figure 7a), based on surveys of random quadrats, and invertebrate orders (Figure 7b) and the order Carabidae (ground beetles) further identified to species level (Figure 7c), with these last two being sampled using pitfall traps located close to the fixed quadrats indicated in Figure 6.

Ordination is essentially a form of factor analysis and the resulting two axes are the leading factors, (i.e. those with the largest eigenvalues, which explain the largest proportions of the data variance). Here the ordinations were performed on a matrix of species information categorized by species type and sample site. The ordination plots (Figure 7) demonstrate the between-patch differences in communities. The ordination plots were created using the detrended correspondence analysis option from the CANOCO computer package (ter Braak, 1988). In the ordination plots, black and white circles represent sample communities from sites A (occasionally flooded) and B (frequently flooded), respectively. Proximity of points is proportional to community similarity, with closely plotting points indicating very similar communities. Ellipses superimposed on the plots encompass 90% of the variance of the ordination scores. Numbered circles represent the locations in the ordinations of particular plant or carabid species, or invertebrate orders. Location of these points relative to the sample points indicates in which samples the orders/species are most frequent and/or abundant. Eigenvalues of the ordination axes are indicated in brackets against the axes labels. Whereas the eigenvlaues for plots (a) and (c) in Figure 7 are of a typical magnitude for ecological data, the low eigenvalues for plot (b) result from the substantial overlap in the taxonomic composition between samples at this level of identification.

The results of the ordination analysis of the vegetation survey data are presented in Figure 7a. Fifty-seven species were recorded. The ordination separates the samples from the A and B sites quite effectively, and also shows greater within-site heterogeneity in site B than in site A (illustrated by the greater spread of sample points and the larger ellipse for site B). The ordination, by separating occasionally and frequently flooded sites, presents a means of examining the association of hydrological and ecological data. Species 4 and 8, placed on the lower left corner of the ordination (at the edge of site A) are typical hedgerow species, whereas species 11 and 12, towards the top right of the ordination (at the edge of site B), are strong indicators of a wetter habitat. The other group of species worth noting are the indicators of a eutrophic environment (species 5, 7 and 9), showing that nutrient status as well as hydrological regime has a strong influence on local ecology.

Both the invertebrate order and carabid species ordinations illustrate considerable heterogeneity of communities within the two sites. In Figure 7b (invertebrate orders), site A samples plot further to the left on the first ordination axis than site B samples. The Hydrophilidae, an invertebrate family that tends to be associated with wet or damp conditions, is shown to be associated with site B. In general, greater within-site heterogeneity is seen for site A than for site B. This characteristic is seen more strongly in Figure 7c, which shows an ordination of carabid beetle species. The distribution of carabids with known preferences for dry or wet conditions also shows that the majority of dry habitat indicators are more frequent and abundant in site A, which is less frequently flooded.

CONCLUSIONS

This paper has illustrated the effectiveness of some simple GIS functions, particularly overlaying, for identifying quite subtle historical changes and contemporary associations between environmental factors that are driven by hydrological processes. The great strength of using a GIS framework for integrating information in this way is the ability to fix locations precisely and to quantify positional error. In this way changes in individual environmental properties and associations between different properties can be identified with confidence.

At the decadal scale, changes in hydrological processes can be seen to cause major changes in the berm extent and morphology, which in turn are providing a new landform upon which vegetation colonization can occur. At a seasonal scale, these areas are characterized by different inundation environments, and support the development of different plant and animal communities.

This paper demonstrates the value of GIS for synthesizing historical and contemporary data to define and map environmental associations to distinguish effectively the functional units along the riparian zone of a regulated river.

ACKNOWLEDGEMENTS

The analysis of historical maps and air photographs was undertaken during the tenure of a Leverhulme Fellowship by one of the authors (A.M.G.). Local topographic survey and ecological research forms part of the ERMAS 2 programme funded by the European Union (DGXII contract ENV4-CT95-0062). Both the Leverhulme Trust and the European Union are very gratefully acknowledged for their support of research on the River Dee. The Environment Agency kindly provided river flow data for the Erbistock/Manley Hall gauging stations.

REFERENCES

Bagnold, R. A. 1966. 'An approach to the sediment transfer problem from general physics'. *United States Geological Survey Professional Paper* No 422, USGS, Washington, D.C.

Bagnold, R. A. 1980. 'An empirical correlation of bedload transport rates in flumes and natural rivers', *Proc. R. Soc. Land. A.*, **372**, 453–473.

Downward, S.R. 1995. 'Information from topographic survey', in Gurnell, A. M. and Petts, G. E. (eds), *Changing River Channels.* John Wiley & Sons, Chichester. pp. 303–323.

Downward, S. R., Gurnell, A. M., and Brookes, A. 1994. 'A methodology for quantifying river planform change using GIS', in: Oliva, L. J., Loughran, R. J., and Kesby, J. A. (eds), *Variability in Stream Erosion and Sediment Transport, IAMS Publ.*, **224**, 449–456.

Gurnell, A. M. 1997a. 'Channel change on the River Dee meanders, 1946–1992, from the analysis of air photographs', *Regul. Rivers, Res. & Mgmt*, **13**, 13–26.

Gurnell, A. M. 1997b. 'Adjustments in channel geometry associated with hydraulic discontinuities across the fluvial-tidal transition', *Earth Surf. Process. Landf.*, **22**, 967–985.

Gurnell, A. M., Downward, S. R., and Jones, R. 1994. 'Channel planform change on the River Dee meanders, 1876–1992', *Regul. Rivers, Res. & Mgmt*, **9**, 187–204.

Harley, J. B. 1975. *Ordnance Survey Maps, A Descriptive Manual.* Ordnance Survey, Southampton.

Lambert, A. 1988. 'Regulation of the River Dee', *Regul. Rivers, Res. & Mgmt*, **2**, 293–308.

Milton, E. J., Gilvear, D. J., and Hooper, I. 1995. 'Investigating change in fluvial systems using remotely-sensed data', in Gurnell, A. M. and Petts, G. E. (eds), *Changing River Channels.* pp. 277–301. John Wiley & Sons, Chichester.

Petts, G. E. 1990. 'The role of ecotones in aquatic landscape management', in Naiman, R. J. and Décamps, H. (eds), *The Ecology and Management of Aquatic-Terrestrial Ecotones*, Man and Biosphere Series. UNESCO Paris and Parthenon, Carnforth. pp. 227–262.

Petts, G. E. and Amoros, C. (eds) 1996. *Fluvial Hydrosystems*, Chapman & Hall, London. 322 pp.

Shi, C., Petts, G. E., and Gurnell, A. M. (submitted). 'Berm development along the regulated lower River Dee, UK',

ter Braak, C. J. F. 1988. 'CANOCO — a FORTRAN programme for canonical community ordination', *Technical Report LWA-88-02 GLW.* Wageningen. 95 pp.

INDEX

Advanced High Resolution Radiometer
(AVHRR) data 151–152, 154
AGNPS model 92
ANSWERS model 92, 95, 96
Applications of GIS
archival 4
coastal sediment plumes and
geography 148–159
DBCP (1,2-dibromo-3-chloropropane)
transport in groundwater 138–144
flood insurance 10
forest harvest effects on peak
streamflow 69–84
groundwater contamination 137–145
hydrological effect of forest roads
80–81
hydrologically-sensitive areas 79–80
land cover change 54
landslides 103–135, 123–134
management 4
modelling 4
natural hazard insurance 10
river morphological and ecological
change 161–173
river planform change 163–169
science 4
soil erosion 85–102
water industry 6–8
water management 1–14
water utilities 6
water yield 54
ARC/INFO GIS 74, 75, 152
Asset management 5
Automated mapping (AM) 6, 9

Berm morphology, inundation frequency,
ecology 169–172

CANOCO statistical package 171
Coastal environment resource
management 147–148
Coastal sediment transfer 150–151
Code of practice 12
COUPLES model 56
Coupling (between GIS and hydrological
models)
close / tight 5, 139
embedded 93–94, 139
loose 5, 91–93, 139

Coupling (between hydrological and slope
stability models) 107–114,
124–125

Data
access 15, 21–22
accuracy 8
error 8, 24, 166
features, attributes, occasions 18–19
meta-data catalogues 22
quality 5, 8
refererence data 20–21
spatially-distributed data availability
70
Data model 7, 12, 15
generic conceptual model 19
water information system (WIS)
18–21
Data policy (UK Natural Environment
Research Council) 17
Data table 20
Database 6, 15–22
relational 16
Decision support 15–22
DEDNM (DEM processing software)
27
Detrended correspondence analysis 171
DH4 overland flow routing algorithm
43–50
DH8 or D8 overland flow routing
algorithm 24, 43–50
Digital contour data 64
Digital Elevation Model (DEM) 23–52,
57, 64, 66, 72, 73, 86,107, 109, 128,
132, 153, 169
causes of depressions and flat areas 25
data errors 39
products derived from 86
raster 23–35
treatment of depressions 23–52
treatment of flat areas 23–35
treatment of upslope contributing
area 37–52
Digital Terrain Model (DTM) 9
Digital topographic data – availability
70
Distributed Hydrology–Soil–Vegetation
Model (DHSVM) 71, 72–82
applications 76–82

calibration 76
land surface characterisation 75–76
topography 74

Elementary spatial unit (ESU) 54
EROSION3D model 93
Error in hydrological modelling 143

Facilities management (FM) 6,8
First order analysis 143
Flow routing 24

Geographical information system (GIS)
138–139
GIS applications see Applications of
GIS
GIS in environmental modelling 139
GIS software
ARC/INFO 74, 75, 152
GRASS 57
IDRISI 108, 110
PCRaster 88, 94
SPANS 165
TOPAZ landscape analysis tool 27,
30
Global Positioning System (GPS) 38,
76, 80
Groundwater contamination and
agriculture 135–136
Groundwater modelling 109–110,
137–145
contamination 137–145
recharge 108–109, 114, 141

HydroGIS 5
Hydrological models
applications 70
applications of spatially-distributed
simulation models 70
applications of spatially-lumped
simulation models 70
distributed 4
error sources 43
parameterization-spatial complexity
54
physically-based 4
topographic parameterization 23
AGNPS 92
ANSWERS 92, 95, 96

COUPLES 56
DH4 43–50
DH8 or D8 24, 43–50
DHSVM 71, 72–82
EROSION3D 93
KINEROS 95
LISEM 94, 98–99
Macaque 54–56
MODFE 109–110
MODFLOW 141–142
MMF 95
MT3D 141–142
PRZM-2 140–142
RHESSys 55, 56
TOPMODEL 55, 56, 57
Topog 56
USLE 86, 91–92, 97, 153
Hydrology – Geographical approach 4

IDRISI GIS 108, 110
Information élites 13
Infrastructure management 6
Inherent error 8
Input error 143
Input–model–output structure 7, 9
Input–transaction–output model 7, 9
Interpolation
   one-dimensional, linear 60–61
   three-dimensional spline 60–61

KINEROS model 95

Land information systems (LIS) 6, 8
Land Ocean Interaction Study (LOIS)
   15–22
Land-Form PROFILE 11
Landsat MultiSpectral Scanner (MSS)
   data 151–152, 154
Landsat Thematic Mapper (TM) images /
   data 62, 151–152, 154
Landslide modelling 107–114, 124–126
   response to / impact of slope toe
     cutting 115, 116
   response to / impact of timber harvest
     114–115, 116–117
   response to incision of channels 115
   temporal patterns of slope response
     117–119
Landslides 103–135

effects of forest management 103–104
hazard assessment 124
impact of root strength 130–131, 133
impacts of shallow landslides 123–124
influences on individual failure
   locations 125–126
large landslide mass-wasting
   processes 103
relationship with critical rainfall
   129–130
relationship with roads 130, 132
size distribution 128–129
topographic control 131
Large-scale spatial models (LSSMs)
   53–54
Leaf area index (LAI) mapping 62–64
LISEM model 94, 98–99

Macaque water balance model 54–56
Management information system (MIS)
   6, 8
Model error 143
Modelling and control systems 8
MODFE model 109–110
MODFLOW model 141–142
Monte Carlo simulation / analysis
   95–96, 143
Morgan/Morgan/Finney model (MMF)
   95
MT3D model 141–142

NOAA GEODAS point bathymetric
   soundings 152
Normalized difference vegetation index
   (NDVI) 62

Object-oriented (OO) programming 22
Operational error 8

Parameter error 143
PCRaster GIS 88, 94
Pesticide leaching mapping 138
Precipitation mapping 58–62
Precision 8
Professional code of practice 12
Professional ethics 13
PRZM-2 model 140–142

Quality assurance (QA) 16

Quality control (QC) 16

Rain-on-snow events 77–79
Regional integrated assessment (RIA)
   144
Remote sensing 6,8, 151–152
RHESSys model 55, 56
Risk assessment 143–144
River margin environments 161–162

Sediment transport capacity index
   89–90
Sensitivity analysis 143
Slope stability modelling 110–114,
   124–126
   Bishop's simplified method of slices
     110–111
   infinite-slope model 124–125
SNOTEL data 76
Soil cohesion 124–125
Soil erosion hazard index 90
Soil erosion risk assessment 86–90
SPANS GIS 165
Spatial resolution 9, 24
Spatial statistics 6, 8
Specific dispersal area (SDA) 43
Specific upslope contributing area
   (SCA) 43
SQL 16
Stream power index 89

TOPAZ landscape analysis tool 27,
   30
   Breaching algorithm 27–30
   Flat area algorithm 30–34
TOPMODEL 55, 56, 57
Topog model 56
Topography mapping 64–67
Total contributing area (TCA) 42–50
Triangulated irregular network (TIN)
   40, 42, 152

Uncertainty characterization 143
Universal Soil Loss Equation (USLE)
   86, 91–92, 97, 153
Upslope contributing area 37–52
USGS digital elevation model 128

Wetness index 88